Lecture Notes in Social Networks

Lecture Notes in Social Networks (LNSN)

Series Editors

For further volumes:
http://www.springer.com/series/8768

Şule Gündüz-Öğüdücü • A. Şima Etaner-Uyar
Editors

Social Networks: Analysis and Case Studies

Springer

Editors
Şule Gündüz-Öğüdücü
Computer and Informatics Department
Istanbul Technical University
Istanbul
Turkey

A. Şima Etaner-Uyar
Computer Engineering Department
Istanbul Technical University
Istanbul
Turkey

ISSN 2190-5428 ISSN 2190-5436 (electronic)
ISBN 978-3-7091-1796-5 ISBN 978-3-7091-1797-2 (eBook)
DOI 10.1007/978-3-7091-1797-2
Springer Wien Heidelberg New York Dordrecht London

Library of Congress Control Number: 2014944729

Printed on acid-free paper

Springer is part of Springer Science+Business Media (www.springer.com)

To Selim and Emre—Ş.G.Ö
To Krie et al.—Ş.E.U

Foreword

Network-based representation and analysis is rapidly growing into an effective technique for data analysis for various applications. The development in the technology directly and tremendously influenced the area of social network analysis and mining by allowing the construction of comprehensive frameworks for social networking and by facilitating the automated analysis of social networks. This has recently attracted considerable attention in the society almost at all levels from practitioners, to researchers, to developers, to end users, etc. Indeed, the amount of research outcome and the number of research groups working on social networks have rapidly increased over the past decades due to the recognition of the social network methodology as an effective model to develop unique solutions for some vital problems. This edited book entitled *Social Networks: Analysis and Case Studies* by Şule Gündüz-Öğüdücü and Şima Etaner-Uyar addresses the need by including a number of chapters written by experts in the field. It aims at helping the reader whether researcher or practitioner to grasp the state of the art in this rapidly growing networking era.

The chapter entitled *Ranking Authors on the Web: A Semantic AuthorRank* by Lule Ahmedi, Lavdim Halilaj, Gëzim Sejdiu, and Labinot Bajraktari introduces the CO-AUTHORONTO as an extended FOAF ontology. The CO-AUTHORONTO is further extended with PageRank and AuthorRank metrics for ranking authors based on their co-author links. The chapter entitled *Detecting Neutral Nodes in a Network of Heterogeneous Agent Based System* by Fatemeh Hendijani Fard and Behrouz H. Far proposes a technique that can identify neutral nodes, i.e., nodes that can never be considered as violating or interesting nodes because they show similar behavior in all of their interactions. David B. Skillicorn and Quan Zheng wrote the chapter entitled *Global Structure in Social Networks with Directed Typed Edges*. The chapter entitled *Social Networks and Group Effectiveness: The Role of External Network Ties* by Fabiola Bertolotti and Maria Rita Tagliaventi explores the relationship between group effectiveness and social networks. Alessia Amelio and Clara Pizzuti wrote a survey about *Overlapping Community Discovery Methods*. Zehra Çataltepe and Abdullah Sönmez address *Classification in Social Networks*. The chapter entitled *Experiences Using BDS: A Crawler for Social Internetworking*

Scenarios by Francesco Buccafurri, Gianluca Lax, Antonino Nocera, and Domenico Ursino argues that a crawling strategy which is good for social networks cannot be expected to remain valid in a Social Internetworking scenario. Accordingly, the authors defined a new crawling strategy specifically conceived for SISs. Lei Chen and Mingxuan Yuan address *Privacy and Ethical Issues in Social Network Analysis*. Finally, the chapter entitled *Social Media: The Evolution of E-health Services* by Tiziana Guzzo, Alessia D'Andrea, Fernando Ferri, and Patrizia Grifoni analyzes e-health services provided by different Social Media and introduces a Hybrid Cloud E-health Services architecture able to provide open, interoperable, scalable, and extensible services for all the e-health activities.

At the end it is important to emphasize this book as a valuable source for people involved in various applications in security, health, multiagent systems, ranking, social network structure and community analysis, among others. The reader will benefit from the content to build a clear vision in the field.

Calgary, AB, Canada Reda Alhajj
March 2014

Preface

With the increasing popularity of Web 2.0, social media has become a widely used communication platform. Parallel to this development, Social Network Analysis gained in importance as a research field, while opening up many opportunities in different application domains. The book is a comprehensive resource for practitioners and researchers alike, who are new to the field as well as those who have years of experience. The book consists of chapters on Social Network Analysis theory and methods with a focus on current applications and case studies applied in various domains such as mobile networks, security, machine learning, and e-health.

This book was conceived as a result of the TUBITAK-EEEAG project 110E027 titled "A Semantic Recommendation System for Social Bookmarking Sites" supported by the Scientific and Technological Research Council of Turkey. We would like to thank our authors for their valuable contributions, without whom this book would not have been possible. They have met our deadlines and then patiently awaited this book to appear in print. We would also like to thank Reda Alhajj for writing the foreword, Tansel Özyer for his continuous support, and Nagehan Ilhan for her help during the preparation of the final manuscript. Finally, special thanks go to the staff at Springer, in particular Annelies Kersbergen.

We hope that you enjoy reading this book.

Istanbul, Turkey
February 2014

Şule Gündüz-Öğüdücü
A. Şima Etaner-Uyar

Glossary

Adjacency matrix A representation of a social network with n nodes using an $n \times n$ matrix where a non-zero entry at position ij indicates the presence of an edge, possible directed, from node i to node j. The magnitude of the entry may be used to represent the strength of the relationship between nodes i and j.

Agent-based system or Multiagent system A system consists of software agents.

AKT A broad ontology, referred to as the AKT Reference Ontology, for describing an academic computer science community. It consists of several sub-ontologies: Support, Portal, etc.

Association for Computing Machinery A premier membership organization for computing professionals. Unlike the IEEE, the ACM is solely dedicated to computing.

Assortativity index A measure of how similar are the labels of connected nodes.

Assortative networks The networks in which nodes with large degree connect to nodes with large degree.

AuthorRank The PageRank algorithm adopted to rank authors.

Blog Blog is an Internet discussion site that consists of discussion posts.

Bridge In a Social Internetworking Scenario, a bridge is a user who has accounts in different social networks and explicitly has specified links among these accounts.

Cliques A clique of a graph G is a complete subgraph of G, and the clique of largest possible size is referred to as a maximum clique. A maximal clique is a clique that cannot be extended by including one more adjacent vertex, meaning it is not a subset of a larger clique. Maximum cliques are therefore maximal cliques (but not necessarily vice versa).

Cloud-Computing Cloud-Computing is a term used to describe different types of computing concepts that involve various computers connected through Internet.

Co-authorOnto An ontology which extends FOAF to support modeling co-authorship networks on the Web.

Co-authorship networks An important class of social networks modeling co-author relations to, say, analyze the trends of research collaborations, or determine the status of individual researchers.

Collaboration diagram Shows the possible interaction path between different types of agents and how they can communicate with each other.

Collaboration matrices Matrix to show the relationships between different agent types containing agents as the nodes and weight of edges between them as the entries of the matrix.

Collective inference Iterative procedures, which classify related instances simultaneously. The content and link information for both training and test data are available. First, based on the available training content, link and label information, models are trained. Then, those models are used to label the test data simultaneously and iteratively where each test sample is labeled based on its neighbors.

Communication Network Network involving all the relationships through which organizational actors share resources such as information, help, and guide related to the execution of their work.

DBLP An on-line repository of bibliographic data on major computer science publications well known to its community.

Degree distribution The degree distribution is the probability distribution of node degrees over a network.

Degrees of nodes The degree of a node is the number of edges connected to the node. In terms of the adjacency matrix A, the degree for a node indexed by i in an undirected network is $k_i = \sum_j a_{i,j}$, where the sum is over all nodes in the network. In a directed network, each node has two degrees. The out-degree is the number of outgoing edges emanating from a node $k_i^{out} = \sum_j a_{j,i}$ and the in-degree is the number of incoming edges onto a node $k_i^{in} = \sum_j a_{i,j}$. The total degree of the node is the sum of its in- and out-degree $k_i^{tot} = k_i^{in} + k_i^{out}$.

Dendrogram A dendrogram is a tree diagram frequently used to illustrate the arrangement of the clusters produced by hierarchical clustering.The top row of nodes represent data (individual observations), and the remaining nodes represent the clusters to which the data belong, with the arrows representing the distance (dissimilarity). The distance between merged clusters is monotone increasing with the level of the merger: the height of each node in the plot is proportional to the value of the intergroup dissimilarity between its two daughters (the top nodes representing individual observations are all plotted at zero height).

Disassortative networks The networks in which large degree nodes tend to connect to small degree nodes.

E-health E-health indicates healthcare practice supported by electronic communication and processes.

Edge prediction Predicting when an edge 'should' be present in a graph, even though it is not, based on the entire structure of the graph.

Edge weight A positive weight associated with the edge in a social network representing the intensity of the relationship between the nodes at its ends.

Edit distance The edit distance between two strings is the number of single character manipulations (insertion, deletion, or replacement) that are needed to transform one string into the other. The edit distance thus describes how similar or dissimilar two strings are by the number of steps it takes to turn from one into the other, where a step is defined as a single character change.

Eigendecomposition A way of transforming the representation of a matrix to use new axes that are aligned along directions of greatest uncorrelated variation. When applied to the representation of a graph, an eigendecomposition is the basis for an embedding into a geometry. This geometry can be used to draw the graph, or to measure the similarity of any pair of its nodes by their distance apart.

Erdős number An early example of co-authorship networks, in which a "collaborative distance," i.e., the smallest number of co-authorship links between any individual author and the mathematician Paul Erdős is calculated.

Friend of a Friend An RDF schema for machine-readable modeling of homepage-like profiles and social networks.

Generalization error The expected test error of a model on any test set that come from the same input distribution as the training set.

Greedy algorithm A greedy algorithm is any algorithm that follows the problem solving metaheuristic of making the locally optimal choice at each stage with the hope of finding the global optimum.

Group effectiveness Multidimensional construct that entails how: groups must produce outputs considered as adequate by those who receive, or who are in charge of evaluating; group members are able to work together in successive tasks; group members derive satisfaction from the execution of their tasks in the group.

Hamming In information theory, the Hamming distance between two strings of equal length is the number of positions at which the corresponding symbols are different. Put another way, it measures the minimum number of substitutions required to change one string into the other, or the number of errors that transformed one string into the other.

Healthcare Healthcare is the prevention and treatment of illness, disease, injury, and other mental and physical impairments in human beings.

Heterogeneous Agent-Based System Agent-based system in which the agents are heterogeneous.

Heterophily The tendency of entities not to be related to other similar entities.

Homophily The tendency of entities to be related to other similar entities.

Indegree The number of times an actor is drawn into an interaction by others.

Institute of Electrical and Electronics Engineers The world's largest professional association dedicated to advancing technological innovation and excellence for the benefit of humanity.

Inter-type relationship Shows the interactions of different agent types.

Intra-type relationship Considers the communications between instances of the same type.

Laplacian matrix A transformation of an adjacency matrix by negating the existing entries and placing the row sums on the diagonal. A Laplacian acts as a kind of normalization, centering the cloud of points corresponding to the nodes around the origin.

Local model The traditional model which use local attributes (features) of the entities for classification.

Me edge It is a link between two accounts of the same user in two different social networks.

Neutral nodes The neutral agents that show the same behavior in all of the possible collaborations. It means that they have a single behavioral pattern when interacting with the other agent types. Therefore the relations between this neutral agent and other agent types may not cause any changes to the network.

Ontology In computer science, it is a formal representation of knowledge in a given domain of interest through vocabulary definition of that domain: the concepts and their taxonomy, properties, constraints, and relationships among concepts.

Outdegree The number of times an actor initiated an interaction.

Outlier An outlier is an observation that lies outside the overall pattern of a distribution.

PageRank An algorithm based on link analysis for ranking. It was initially used for ranking Web pages on the Web by Google, but is also nowadays used to rank authors (AuthorRank), detect spams, and similar. There is a metric named PageRank in social network analysis (SNA) as well, which measures the reputation of an individual in a network following the same rationale as when ranking Web pages.

Random Graph A random graph is a graph in which properties such as the number of graph vertices, graph edges, and connections between them are determined in some random way.

Random Walk matrix A transformation of an adjacency matrix obtained by dividing each entry by the corresponding row sum. The entries can then be regarded as the probabilities of leaving a node on each of its outgoing edges during a random walk of the graph.

Random walk A random walk is a route consisting of successive and connected steps in which each step is chosen by a random mechanism uninfluenced by any previous step.

Relational model The model that uses the relations in the network as well as the values of attributes of related entities for classification.

Resource Description Framework The basic standard for the Semantic Web. It defines the building blocks of the Semantic Web, such as classes and properties, and how they interact to create meaning.

Semantic Protocol and RDF Query Language The standard language for querying RDF data.

Semantic Query-Enhanced Web Rule Language A SWRL-based language for querying OWL ontologies.

Semantic Web Rule Language A widely used language to express rules in the Semantic Web.

Semantic Web As opposed to the traditional Web, it represents the Web of data with meaning in such a way that a computer program can understand (make use of) that meaning of the data.

Semi-supervised classification A classification method which makes use of test nodes or some unlabeled nodes during training.

Single-linkage Single-linkage algorithm is an example of agglomerative hierarchical clustering method. We recall that is a bottom-up strategy: compare each point with each point. Each object is placed in a separate cluster, and at each step we merge the closest pair of clusters, until certain termination conditions are satisfied. This requires defining a notion of cluster proximity. For the single-linkage, the proximity of two clusters is defined as the minimum of the distance between any two points in the two clusters.

Social Internetworking Scenario It is a set of different social networks which, thanks to the presence of common users, allows users to interact with each other even if they join different social networks.

Social Media Refers to the means of interactions in which people share, create, and exchange information in virtual communities.

Social Network Analysis An interdisciplinary area of research in social sciences and the information technology. It provides theories and techniques that prove the effects of an individual or a group of individuals belonging to a given network into some outcomes related to that individual or group.

Social Networking Social Networking is group of people that discuss and share ideas and/or specific interests on the Internet.

Software agents Knowledgeable, autonomous, situated, and interactive software entities. They can have other characteristics as well.

Spectral embedding A family of techniques for embedding a graph in a geometry based on its eigenstructure.

Transductive learning Semi-supervised learning where the identity of the test nodes are known and therefore the learning method only needs to provide the outputs on the test nodes and not necessarily a learned function.

Uniform Resources Identifiers Initially used as a standard format to identify pages on the Web (a string usually starting with http://). Today it serves to identify simply any thing of interest, be it physical or conceptual, accessible by augmenting it to the Internet. Semantic Web also uses http URIs as identifiers for its resources.

Web Ontology Language The standard language to formally define and instantiate ontologies on the Web so that they can be used and understood in that certain domain.

Work group Bounded set of individuals, working interdependently towards a common goal.

Workplace Social Networks Ongoing social relationships in which organizational actors are embedded in the workplace.

Contents

List of Contributors

Lule Ahmedi Faculty of Electrical and Computer Engineering, University of Prishtina, Prishtina, Republic of Kosovo

Alessia Amelio National Research Council of Italy (CNR), Institute for High Performance Computing and Networking (ICAR), Rende, Italy

Labinot Bajraktari Faculty of Electrical and Computer Engineering, University of Prishtina, Prishtina, Republic of Kosovo

Fabiola Bertolotti University of Modena and Reggio Emilia, Reggio Emilia, Italy

Francesco Buccafurri DIIES, University Mediterranea of Reggio Calabria, Reggio Calabria, Italy

Lei Chen The Hong Kong University of Science and Technology, Clear Water Bay, New Territories, Hong Kong

Zehra Çataltepe Computer Engineering Department, Istanbul Technical University, Istanbul, Turkey

Alessia D'Andrea IRPPS-CNR, Rome, Italy

A. Şima Etaner-Uyar Computer Engineering Department, Istanbul Technical University, Istanbul, Turkey

Behrouz H. Far Department of Electrical and Computer Engineering, University of Calgary, Calgary, AB, Canada

Fernando Ferri IRPPS-CNR, Rome, Italy

Patrizia Grifoni IRPPS-CNR, Rome, Italy

Şule Gündüz-Öğüdücü Computer Engineering Department, Istanbul Technical University, Istanbul, Turkey

Tiziana Guzzo IRPPS-CNR, Rome, Italy

Lavdim Halilaj Faculty of Electrical and Computer Engineering, University of Prishtina, Prishtina, Republic of Kosovo

Fatemeh Hendijani Fard, Department of Electrical and Computer Engineering, University of Calgary, Calgary, AB, Canada

Nagehan İlhan Computer Engineering Department, Istanbul Technical University, Istanbul, Turkey

Gianluca Lax DIIES, University Mediterranea of Reggio Calabria, Reggio Calabria, Italy

Antonino Nocera DIIES, University Mediterranea of Reggio Calabria, Reggio Calabria, Italy

Clara Pizzuti National Research Council of Italy (CNR), Institute for High Performance Computing and Networking (ICAR), Rende, Italy

Gëzim Sejdiu Faculty of Electrical and Computer Engineering, University of Prishtina, Prishtina, Republic of Kosovo

David B. Skillicorn School of Computing, Queen's University, Kingston, ON, Canada

Abdullah Sönmez Computer Engineering Department, Istanbul Technical University, Istanbul, Turkey

Maria Rita Tagliaventi Department of Management, University of Bologna, Bologna, Italy

Domenico Ursino DIIES, University Mediterranea of Reggio Calabria, Reggio Calabria, Italy

Mingxuan Yuan Hong Kong Huawei Noah's Ark Lab, Hong Kong

Quan Zheng School of Computing, Queen's University, Kingston, ON, Canada

Introduction to Social Networks: Analysis and Case Studies

Nagehan İlhan, Şule Gündüz-Öğüdücü, and A. Şima Etaner-Uyar

Abstract Social networks are platforms to share media, ideas, news, links or any kind of content between users and their neighbors, thus providing a perfect reflection of the structure and dynamics of the society. The recent advances in social media and the growing use of social networking tools have lead to explosive growth of information available over the Internet and created a need to better understand the underlying structure of the knowledge flow. Social Network Analysis focuses on analyzing the relationships within and between users/groups in order to model the interactions and includes assumptions about how best to describe and explain the social network. Social Network Analysis and understanding the dynamics of social networks have become popular research topics and a vast number of studies have been performed. This chapter provides definitions of the basic concepts of Social Network Analysis and briefly introduces the topics covered in the book.

1 Introduction

This chapter will provide an introduction to social networks and give a description of resources and principal topics covered by Social Network Analysis (SNA). A social network is a social structure made up of actors called nodes, which are connected by various types of relationships. SNA is used to analyze and measure these relationships between people, groups and other information/knowledge processing entities and provides both a structural and a mathematical analysis. Therefore, the objects under observation are not actors and their attributes, but the relationships between actors and their structure. Relationships show what kinds of information are exchanged between which actors. SNA techniques are used to identify the

N. İlhan (✉) • Ş. Gündüz-Öğüdücü • A.Ş. Etaner-Uyar
Computer Engineering Department, Istanbul Technical University, Istanbul, Turkey
e-mail: nilhan@itu.edu.tr; sunduz@itu.edu.tr; etaner@itu.edu.tr

Ş. Gündüz-Öğüdücü and A.Ş. Etaner-Uyar (eds.), *Social Networks: Analysis and Case Studies*, Lecture Notes in Social Networks, DOI 10.1007/978-3-7091-1797-2_1,

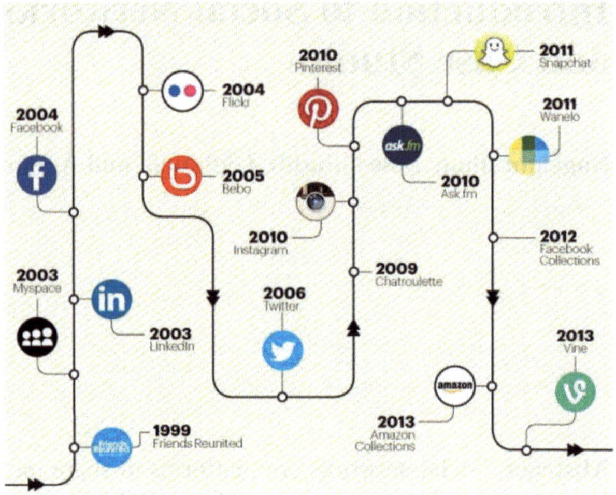

Fig. 1 History of online social networks [43]

characteristics of positions held by the actors in a social graph and the characteristics of the graph structure. With the advent of social network based applications, there has been more interest in computational techniques to discover the properties of networks. SNA has also attracted a significant interest in many fields such as sociology, epidemiology, anthropology, social psychology, communication studies, information sciences, etc.

Social networking has the ability to connect geographically dispersed users and provides social contact using the Internet. Social networks have become very popular in recent years due to the increasing usage of Internet enabled devices such as personal computers and mobile devices. The ever increasing popularity of many online social networks such as e.g. Facebook,[1] Twitter,[2] MySpace[3] etc. is a good indicator for this observation. Figure 1 illustrates the history of online social networking sites in terms of when they were created. However, the social network concept is not restricted to internet based social networks.

Previous studies on SNA generally do not focus on online social interactions. In the last century, researchers in the behavioral sciences have stated studying social networks of offline interactions, such as person to person interactions, letters, telephone conversations, and so on. According to John Scott [42], there are three main research lines in SNA:

- Sociometric Analysts used graph theory and methods developed by Jacob Moreno who first introduced the concept of a sociogram [28]. A sociogram is

---

[1]http://www.facebook.com/.

[2]http://www.twitter.com/.

[3]http://www.myspace.com/.

a visual diagram of relationship networks in which individuals are represented as points and their links to others as lines.
- Harvard University researchers first studied clique formations in social groups to identify cohesive subgroups in social systems in 1930s [24, 46].
- A group of anthropologists in Manchester University studied relational structures characterizing a tribal community. They directed their attention to people's informal social relationships rather than those associated with institutions and associations. Their work focused on the effective configuration of relationships deriving from conflicts between individuals and changes in these networks. John Barnes introduced the term 'Social Networks' and provided a remarkable advancement in the analysis of social structures [5].

Inspired by these studies, Harrison C. White and his colleagues focused on exploring the mathematical basis of social structures. They introduced important algebraic characteristics through the use of algebraic models of groups based on set theory, aiming to formalize different structural relations inside a group. The main idea of this study is that the search of structures in a network should be based on real interactions between the nodes and on how these interactions affect it, instead of on categories defined in advance. Later on, Mark Granovetter proposed a study on the importance of weak ties called "the strength of weak ties". He claimed that weak ties can be important in seeking information and members of a clique should look beyond the clique to their other friends and acquaintances in order to find new information [17]. In addition, a novel theory known as the small world phenomenon was proposed by Stanley Milgram [26]. In his famous small world experiment, a sample of individuals were asked to reach a particular target person by passing a message along a chain of acquaintances. The median length of the successful chains turned out to be five intermediaries or six separation steps. These studies constituted the foundation of the methods in SNA today.

The rest of the chapter is organized as follows. In the next section, we provide general definitions in SNA. In Sect. 3, tools commonly used to manipulate social networks are given. In Sect. 4, we discuss the remaining chapters in this book. Section 5 concludes the chapter.

2 Definitions in Social Network Analysis

2.1 Graphs

Social networks are usually represented as graphs. A graph $G(V, E)$ consists of a set of nodes V and a set of edges E.

- A graph may be directed or undirected: for instance, an e-mail may be from one person to another and will have a directed edge, or a mutual e-mailing event may be represented as an undirected edge.

- Graphs may be weighted; there may be multiple edges between two nodes. In a weighted graph G, let $e_{i,j}$ be the edge between node i and node j and $w_{i,j}$ is the weight of $e_{i,j}$. The total weight w_i of node i is the sum of weights of all its neighboring edges:

$$w_i = \sum_{n=1}^{d_i} w_{i,n} \tag{1}$$

  where d_i represents its degree.
- Graphs may be unipartite, bipartite or multipartite.
  - A unipartite graph is a normal graph whose nodes are individuals and links. Nodes belong to the same class.
  - Bipartite graphs are graphs where nodes belong to two classes with no edges between nodes of the same class.
  - Multipartite graphs are graphs where nodes belong to more than one class, with no edges between the nodes of the same class.

A graph is generally described by an adjacency matrix A which is a square matrix with as many rows and columns as the nodes in the graph. The $[i, j]$ element of the adjacency matrix corresponds to the information about the ties between nodes i and j. An adjacency matrix may be symmetric (undirected graphs) or asymmetric (directed graphs). All entries of the adjacency matrix of an unweighted graph are 0 or 1 where a zero indicates that there is no tie and a one indicates that a tie is present between the nodes. For a weighted graph, the value of the $[i, j]$ entry in the matrix is the weight that labels the edge from node i to node j.

2.2 *Fundamental Metrics*

The study of SNA involves the measurement of particular structural metrics in order to understand the fundamental concepts of social graphs. Metrics are used to characterize and analyze connections within a given social network. Some of these metrics represent the characteristics of individual nodes whereas others infer a pattern that belongs to the network as a whole. Here, we describe the fundamental metrics that are used in SNA.

Centrality: Centrality measures the relative importance of a node and gives an indication about how influential a node is within the network. Betweenness Centrality, Closeness Centrality, Degree Centrality and Eigenvector Centrality are all measures of centrality.

Betweenness Centrality: Betweenness counts the number of shortest paths in a network that passes through a node. It takes into account the connectivity of the node's neighbors by giving a higher value for nodes which bridge clusters.

The minimum number of edges that must be traversed to travel from a node to another node of a graph is called the shortest path length. Nodes having a high betweenness value play an important role in communication within the network [12]. Betweenness centrality C_B for a node i is calculated as:

$$C_B(i) = \sum_{j<k} g_{jk}(i)/g_{jk} \tag{2}$$

where $g_{jk}(i)$ is the number of shortest paths between j and k that pass through i and g_{jk} represents all paths between j and k.

Closeness Centrality: Closeness centrality measures how close a node is to all the other nodes. A node is considered important if it is relatively close to all other nodes. Closeness is based on the inverse of the distance of each node to other nodes in the network [12]. Closeness centrality C_C for a node i is calculated as:

$$C_C(i) = [\sum_{j=1}^{N} d(i,j)]^{-1} \tag{3}$$

where $d(i,j)$ is the geodesic distance between node i and j. The geodesic distance is the length of the shortest path between two connected nodes.

Degree Centrality: Degree centrality of a node is the number of links to other nodes in the network. A node's in or out degree is the number of links that lead into or out of the node, respectively. In an undirected graph they are identical. This measure can be used for evaluating which nodes are central with respect to transferring information and influencing others in their immediate neighborhood [31]. It can be calculated by using the adjacency matrix:

$$C_D(i) = \sum_{j=1}^{n} a_{ji} \tag{4}$$

where a_{ji} is the $[j,i]$ entry of the matrix.

Eigenvector Centrality: A node's eigenvector centrality is proportional to the sum of the eigenvector centralities of all nodes directly connected to it [6]. It is a useful measure in determining which node is connected to the most connected nodes. For a node i, the Eigenvector Centrality is defined as:

$$C_E(i) = v_i = 1/\lambda_{max}(A) \sum_{j=1}^{N} a_{ji} v_j \tag{5}$$

where $v = (v_1, \ldots v_n)^T$ is the eigenvector for the maximum eigenvalue $\lambda_{max}(A)$ of the adjacency matrix A.

Clustering Coefficient: Clustering coefficient is a measure of the degree to which the nodes in a graph tend to cluster together. The clustering coefficient of a node i is the fraction of pairs of i's neighbor nodes that are connected to each other by edges [48]. Clustering coefficient for node i with degree $d(i) \geq 2$ is defined as:

$$C_{cc}(i) = \delta(i)/\tau(i) \tag{6}$$

where $\delta(i)$ is the number of three connected cliques defined as $\delta(i) = \{\{i, j\} \in E : \{\{j, k\} \in E : \{\{i, k\} \in E\}$ and $\tau(i)$ is the number of triples of node i. Triple of a node i is a path of length two for which i is the center node.

Density: Density is the ratio of the number of edges in the network over the total number of possible edges between all pairs of nodes. It is a useful measure in comparing networks against each other. Density of a graph is calculated as follows:

$$G_G = 2 * |E|/(|V| * (|V| - 1)) \tag{7}$$

where $|V|$ is the number of nodes and $|E|$ is the number of edges in the network.

Path Length: Path length is a measure of the distances between pairs of nodes in the network. Average path length in a network is the average of these distances between all pairs of nodes. A shorter average path length means that the information will spread faster within the network.

Radiality: Radiality shows how far a node reaches into the network. It also measures the amount of novel information provided by the node and the influence it induces on the network. Nodes that have high radiality values usually have convenient positions to be innovators, thus they can relay the ideas in their neighborhood into other parts of the network.

Structural Cohesion: Structural cohesion is defined as the minimal number of nodes in a social network that need to be removed to disconnect the group [27]. It is used to identify cohesive subgroups in a network and reveals how such groups relate to one another.

3 Social Network Analysis Tools

SNA tools are used to represent, analyze and simulate a network by describing the features of the network either through a numerical or a visual representation. Network analysis tools enable researchers to examine different sizes of networks, from small to very large. Representation, visualization, characterization and community detection are expected functionalities of an analysis software. A software should be able to represent both the directed and the undirected structure of a network. Visualization of social networks is also important to understand the network data and examine the result of the analysis. It allows to display nodes and edges in various layouts and distinct colors, size and other advanced properties of the network representation. Many quantitative measures have been defined to characterize networks. Computation of various measures provided at the node level or on the whole graph should be done by a software. Communities are groups of nodes, where nodes within the same community tend to be highly similar, sharing common features, while nodes of different communities have low similarity. Detecting and evaluating the community structure of graphs constitutes an essential task in SNA. Thus, community detection is one of the expected functionalities of an analysis software.

A wide variety of tools, each specialized on one or more of the expected functionalities, exist. Most prominent tools can be listed as follows:

1. **Gephi** Gephi provides an open source software package and a Java library for graph and network analysis [14]. It is a graph visualization tool that uses a 3D render engine to display large networks and works with huge datasets and graphs. Gephi is very convenient to explore dynamic networks. It supports graphs whose content changes over time, and has a timeline component where a timestamp of the network can be retrieved. From the time range of the timestamp, the software retrieves all nodes and edges that match and update the visualization module. Therefore, it allows to display a dynamic graph as a movie sequences. Gephi's framework offers the most common metrics like betweenness, closeness, diameter, clustering coefficient, average shortest path, PageRank, HITS, community detection using modularity, random generators for SNA.
2. **Pajek** Pajek is a network analysis package that performs analysis and visualization of large networks. The main goals in the design of Pajek are decomposition of a large network into several smaller networks, providing powerful visualization tools and implementing subquadratic algorithms to analyze large networks. Pajek analysis and visualization are performed using different data types like graph, partition, vector, cluster, permutation and general tree structures.
3. **NetworkX** NetworkX is a Python library for network analysis [18]. It is used for the creation, manipulation, and study of the structure, dynamics, and functions of complex networks. The library includes support for reading and writing graphs in various file formats. It also includes functions to generate graphs according to a variety of well known graph generation models. NetworkX is well documented, but the clustering algorithms are missing. NetworkX is not primarily a graph drawing package but basic drawing with Matplotlib as well as an interface to use the open source Graphviz software package are included. For advanced visualization, other tools should be preferred.
4. **Igraph** Igraph is a library for network analysis which uses Python and the R environment. It is one of the most essential libraries and is used in large graphs, similar to NetworkX [19]. It also provides properties for graph statistics such as computing degree centrality, closeness centrality and betweenness centrality etc. Dyad and triad census are available in Igraph and Pajek. Igraph offers a

few community detection algorithms like Walktrap [36], Fast Greedy [16], Label Propagation [37], etc.
5. **JUNG** JUNG(the Java Universal Network/Graph Framework) is an open-source software JAVA library which is mainly developed for creating interactive graphs [32]. Main functionality of JUNG is network and graph manipulation, modeling, analysis, and visualization. It has built-in support for GraphML, Pajek, and some text formats. It presents customizable visualization, graph types and includes graph theory, data mining and SNA algorithms (random graph generation, clustering, decomposition, optimization, statistical analysis, distances, flows, and centrality measures). It supports a native sparse matrix format and a graphical user interface, which makes JUNG's representations and algorithms both space and time efficient.

4 Topics in Social Network Analysis

The SNA field encapsulates a wide range of research topics and new methods and approaches are continuously being developed. Therefore, it is very hard to cover the entire network analysis literature. In this section, the topics are covered within the scope of the chapters in the book.

4.1 Node Analysis

The emergence of social networks has resulted in an exponential increase in the amount of the information about individuals, their activities, connections and features representing the characteristics of the individuals. These features may be of different types: demographic features like age, gender; features which represent political and religious beliefs; features representing hobbies, interests, affiliations etc. These features appear on the user's profile within the network, or attached to other objects like photos or videos. There are many applications that can make use of these features and connections like suggesting new connections to individuals based on finding individuals with similar interests and relationships, recommendation systems to suggest movie, music or other products, advertising systems which show advertisements to individuals in which they will most likely be interested. Thus, analyzing the nodes from different perspectives such as how closely related individuals are, who is the connector or hub in the network, who has best visibility of what is happening in the network, what are the distances and similarities of individuals from each other is crucial.

Widely used measures to identify influential nodes within the network are the degree centrality, closeness centrality and betweenness centrality [12]. However, some other methods like PageRank is adopted to find effective nodes. PageRank is a popular Google patented algorithm to examine the entire link structure of the Web and determine which pages are most important by viewing the hyperlinks as recommendations [33]. A page with more inlinks (recommendations) must be important than a page with a few inlinks. A Web page is important if it is hyperlinked by an important recommender. In chapter "Ranking Authors on the Web: A Semantic AuthorRank", a model is proposed to rank authors on the Web like ranking Web pages by considering their co-author links. They have adopted FOAF (friend of friend ontology), the so-called CO-AUTHORONTO ontology, in order to represent authors and also their co-author links on the Web. CO-AUTHORONTO extended with PageRank and AuthorRank metrics for ranking authors is based on the co-author links of the authors. Their framework builds on top of several known ranking metrics and side parameters like the number of authors, co-authorship exclusivity, PageRank, and co-author weight.

In chapter *Detecting Neutral Nodes in a Network of Heterogeneous Agent Based System*, a method is proposed to detect neutral agents in a social network using multi agent systems. They aim to reduce the complexity of analysis in a network consisting of heterogeneous software agents as nodes. Their method suggests detecting neutral agents (the agents that behave with similar frequencies and have similar behaviour) with respect to inter-type and intra-type communications. Thus, the complexity of the network and the algorithms to analyze them will be decreased by identifying and eliminating these agents.

4.2 Edge Analysis

Social networks are usually modeled using graphs where an edge between two nodes represents a relationship between them. Every node and each of the corresponding edges belonging to the nodes carry certain characteristics. Each node represents an entity, while each edge carries attributes that describe the nature of the relationship. There exist two types of social networks: homogeneous and heterogeneous [7]. In homogeneous social networks, there is only one kind of relationship between nodes and the knowledge flow is through this relationship. Heterogeneous social networks have several kinds of relationships between nodes and are also known as multi-relational social networks. Thus, the knowledge flow is through different kinds of relationships and network elements exchange different types of knowledge according to the type of the relationship. In the real world, social networks are usually multi-relational and users establish a large number of relationships with varying edge strengths and types: friends, family, colleagues and so on. Each relation defines a single relational network. A multi-relational network can be defined as a merger of multiple single relational networks. An edge between two nodes consists of all relations and interactions between the two individuals [9]. The weight of an edge in a multi-relational social network should consider the weight of all relationship types between the two nodes. Analyzing multi-relational networks is much more complex than a single network analysis. In single relational networks, nodes between different groups are assumed to be loosely connected but in multi-relational networks the edges between different groups may be as dense as the edges within the groups. The authors in chapter *Global Structure in Social Networks with Directed Typed Edges*, present a spectral approach to understanding and analyzing graphs with a finite number of edge types. Their contribution is to extend conventional spectral graph analysis to networks where edges have different types. In this manner, it is possible to combine features of different types of social networks into a single network framework by taking into account the qualitative differences in the functionality of edges. They construct a multi-layer graph and embed this multi-layer graph using a directed-graph spectral technique. Their technique enables to answer several edge prediction questions such as if there should be an edge between given two nodes and what type of edge is it likely to be.

In a social graph, the presence of an edge between any two nodes indicates the efficiency of communication between them. It also means that they may belong to the same social group and work together. Understanding the structure and dynamics of social groups is crucial for network analysis. From the organizational perspective, groups are core organizational work units. The effectiveness of a group can provide a large contribution to the organizational success. Social edges in work groups are informal links between group members. Group members have different skills and capabilities which are essential for the effectiveness of the group and thus for the organization as a whole. Interaction within a group or interaction between groups are also important factors in a group's processes and outcomes. In literature, many studies exist that analyze the factors contributing to team effectiveness [20, 21, 41]. Unresolved empirical questions exist about the correlation between group density and group effectiveness. Studies show that social interactions and the communication frequency between members of a group are positively related with team effectiveness [4, 39]. They state that teams with densely configured interpersonal edges reach their goals better. In chapter *Social Networks and Group Effectiveness: The Role of External Network Ties*, the relationship between group effectiveness and social networks is examined. A communication network is formed using a 5-month ethnographic observation within three work groups employed in an Italian clothing company by recording all interactions occurring within the groups and between the groups. They show how qualitative information on the nature and dynamics of the ties between group members and other organizational actors can enhance comprehension of the impact of network relationships on organizational behaviors. They claim that the prolonged observation of group members' interactions offers researchers a privileged, thorough perspective into a group's social network. They emphasize that a high centrality degree in the request for information/advice network as opposed to the reporting of a problem or the communication of information/advice network can generate different effects on members' behaviors and on the evaluation of groups' effectiveness. In addition, they highlight the positive outcomes of team leaders who have also a high external prestige, in addition to internal prestige.

4.3 Community Detection and Classification

Extraction of social communities is one of the most important problems in today's SNA. Community detection attempts to solve the identification of groups of vertices that are more densely connected to each other than to the rest of the network. Communities correspond to groups of nodes in a graph that share common properties or have a common role in the organization of the system [16]. The ability to find and analyze communities has proved invaluable in understanding the underlying structure of the network. A number of methods to address this problem have been proposed. They vary in the type of network they can handle (unpartite vs. bipartite, weighted vs. unweighted, etc.) and the type of community structure they can detect (disjoint, overlapping, hierarchical, etc.). A comprehensive recent survey of community detection algorithms is proposed in [11]. The vast majority of the community detection algorithms find disjoint communities. However, in real networks, communities are usually overlapping which means that some nodes may belong to more than one community. Thinking about a person's personal social network, it is naturally considerable that a person may belong to several communities: for example family, co-workers, college friends, and so on. These kinds of networks are usually defined as overlapping networks [34]. However, most of the community detection algorithms find discrete communities which do not capture the overlapping community structure. Thus, for a correct representation of real network communities, it is crucial to find overlapping community structures. In chapter *Overlapping Community Discovery Methods: A Survey*, a review of the most recent proposals in the topic of overlapping community detection is introduced. Methods are classified by taking into account the underlying principles guiding them to obtain division of a network into groups sharing part of their nodes.

With the emergence of social networks, a large amount of information about individuals, their activities, connections has become available. A large part of these individuals which are represented as nodes in a graph structure may be labeled. This leads to a problem of providing correct and high quality labeling for every node, in other words, the node classification problem. In literature, there are a variety of node classification techniques. The simplest one is to use data about the labeled nodes and use a simple classifier in order to classify the unlabeled ones based on only those attributes that are local to these nodes. The techniques that work in this manner are called local classifiers. As in classical machine learning techniques, first, the features of the nodes should be identified. These features may be properties common to all nodes like age, gender, homeland, etc. But the existence of the connections and relationships makes the graph labeling problem different from the traditional machine learning classification techniques, where the classified nodes are assumed to be independent. The techniques that use the link information of the graph are called relational classifiers. Additional features such as the degree, centrality and so on based on adjacency in the graph can be defined in order to achieve a higher classification accuracy. Also, the labels of the neighbors constitute a useful feature. In social networks, the edges indicate some degree of similarity between the connected nodes and constitute a useful input for the learning algorithm. Homophily and co-citation regularity are the two important phenomena used in the labeling process of the nodes. The labeling process can be iterative. Iterative algorithms use local neighborhood information to generate features that are used to learn local classifiers [30]. An iterative algorithm assumes that all of the neighbors' attributes and labels of that node are already known in determining the label of a node. Then, it calculates the most likely label with a local classifier which uses node content and neighbors' labels. In chapter *Classification in Social Networks*, the details of a method which uses content, link and label information on social network data for classification are given. Important properties of social network data which may be used to characterize a social network dataset and a list of aggregation operators which are used to aggregate the labels of neighbors are defined. A number of label aggregation methods are also experimented with. Different classifier accuracies with usage of only the content, only the link or both the content and the link information are evaluated. It is shown that homophily plays an important role in evaluating whether the network information would help in classification accuracy or not.

4.4 Graph Crawling

With the emergence and popularity of social networking sites, such as Facebook, Linkedin, MySpace, Twitter, etc., the number of users joining these sites has dramatically increased. Alexa [2], a well-known traffic analytics website, reported that Facebook is the second most visited website on the Internet, Linkedin ranked as the eighth and Twitter follows them with the tenth rank. Thus, online social networks have become an important phenomenon on the Internet. This global phenomenon has generated lots of interest in many disciplines to analyze human social behavior depending on observations of these networks. However, it is usually not possible to obtain datasets from the Online Social Network services due to privacy issues and it is hard to get such data directly from the service providers. Moreover, the huge size and access limitations of most of the services make it hard to completely cover the whole social graph. A widespread approach is crawling and sampling the network. It is desirable to crawl a small but representative sample of the network. In crawling, a user is randomly chosen and the friend list of the user is retrieved. Again one user is selected from out of the friend list and a new list of friends retrieved. In principle, the process repeats until every user in the network has been visited.

Several crawling strategies for single social networks have been proposed. These methods differ in the selection of the next friend. Breadth First Search (BFS) [50], Depth First Search (DFS) [47], Simple Random Walk (SRW) [23], Simple Random Walk with re-weighting (SRW-rw) [38,40] and Metropolis-Hastings Random Walk (MHRW) [25] are the most popular ones among them. In BFS and DFS techniques, the graph is crawled node per node adding all discovered nodes to a list of nodes to visit. But, the difference between BFS and DFS techniques for graph crawling is the order in which the next node in the graph is selected. BFS selects the first node of the list as the next node to visit and removes it from the list, while DFS selects the last node and marks it as visited. In Simple Random Walk technique, the next node is chosen uniformly at random among the neighbors of the current node. This algorithm is biased towards the high degree nodes. SWR-rw operates based upon a sequence of random nodes obtained by a Simple Random Walk with a proper re-weighting process to provide the unbiased sampling. MHRW is a crawling technique which applies the Markov Chain Monte Carlo (MCMC) method [15] for sampling from a probability distribution that is difficult to sample from directly. SWR-rw and MHRW ensure unbiased graph sampling. These crawling techniques are good for single social networks but they are not suitable for Social Internetworking Systems, where users are members of multiple social networks.

Recent studies show that users are often affiliated with different social network sites and in each one of them they exhibit different site-specific interaction patterns. This knowledge provides better understanding of the users' tastes and improves the quality of service they can get. The authors in chapter *Experiences Using BDS: A Crawler for Social Internetworking Scenarios*, confirm that a crawling strategy which is good for single social networks should not be expected to be appropriate for SIS, due to their specific topological features. They also propose a new crawling strategy Bridge-Driven Search (BDS), specifically designed for SISs, which overcomes the drawbacks of other crawling strategies. BDS is based on the bridge concept, which represents the structural element that interconnects different social networks. Bridge nodes are described as the users who joined more than one social network and explicitly declared their different accounts. They conduct several experiments and show that BDS outperforms state-of-the-art techniques. In addition they perform a large number of experiments to derive detailed information about the bridge nodes and argue that most of the required information on the structural properties of SISs can be obtained through studying bridges in detail.

4.5 Privacy and Social Networking Ethics

Online social networks such as Facebook, MySpace, Twitter, Orkut, Linkedin and etc. are Web sites which are used widely to build connections and relationships. In principle, they allow their users to communicate with other people around the world, contact their closest friends, share experiences, photos, videos with them in real time. Users publish detailed personal information and information about their preferences and daily life. Social Web sites are also collecting a variety of data about their users, both to personalize the services for the users and to sell them to advertisers. However, some of the data revealed in these networks should remain private and not be published at all. Besides, scammers, identity thieves, stalkers, and companies looking for a market advantage are using social networks to gather information about customers.

In literature, some key concepts of privacy on online social networks are identified [35]. These concepts are network anonymization [22, 29], privacy preservation

[1, 45] and access control [8, 10]. Privacy issues of social networks include the disclosure of nodes' identity information, relationship information, data information related to nodes, etc. Network anonymity graphs are obtained by removing the nodes' identity information in social networks in order to preserve privacy. This makes node identification difficult for attackers. However, anonymization is not sufficient to protect privacy and the attackers still identify nodes' identity from the features and the structure of the network. Privacy preservation focuses on protecting sensitive information of users through techniques based on hiding sensitive attributes, identities and modifying data. But hidden attributes of the users can still be inferred. For example, it is possible to predict the home address of the user by analyzing the geographical place of the most frequent updates posted or it is possible to predict the work address by analyzing the relationships, namely if the majority of the users' friends are in the same institution, the user is most likely to work there. Access control mechanisms are used to reinforce access to users' sensitive information without explicit authorization by performing appropriate access control mechanisms. Many existing social network owners offer access control mechanisms that are primitive, permitting coarse-grained visibility control to users to place restrictions on who may view their personal information. Indeed, even with current access control mechanisms, users loose their control over data after its very first publication in the network. From the service owner perspective, it is crucial to protect users privacy while providing useful data. In chapter *Privacy and Ethical Issues in Social Network Analysis*, privacy issues in social network data have been discussed. Different aspects of graph publication issues in graph publishing models have been introduced.

4.6 Cloud Computing with Social Media

Cloud computing is the delivery of computing services over a network such as the Internet. It allows individuals immediate access to a large number of supercomputers and their corresponding processing power that exist at remote locations. The cloud computing model allows users to access information and computer resources from anywhere with an available Internet connection. Social networking sites, online file storage, webmail, online software applications are examples of cloud services. Cloud computing service models can be divided into three main categories: Software as a Service, Platform as a Service and Infrastructure as a Service. Software as a Service (SaaS) provides running applications without installing on your hard drive and without any configuration requirements. Applications are hosted by a service provider and are made available to users over Internet. Platform as a Service (PaaS) model provides a development and production environment which consists of the operating system, the hardware and the network. Users install or develop their own software and applications.

The Infrastructure as a Service (IaaS) model provides the use of resources such as virtual machines, storage etc. It supplies the hardware and network resources; the user installs or develops its own operating systems, software and applications. Cloud computing services reduce the cost and complexity of owning hardware, installing and configuring infrastructure. The other benefits to users are scalability and reliability. Cloud computing services are scalable because they scale up and down as needed without any cost and they offer processing and storage capacity. They are reliable since applications, documents and data are accessible anywhere in the world via the Internet [13].

Cloud computing services are integrated within online social networking sites in a variety of forms. Typically, cloud platforms are used to host social networks or scalable applications being created and hosted in the cloud. Cloud computing drives many social networking sites that has been accessed virtually by millions of users every day. When the user stores photographs to Flickr, posts them to Facebook or uploads a video to YouTube, the corresponding media are stored in the cloud. For instance, Facebook which is the most popular social networking site, provides scalable cloud based applications hosted by Amazon Web Services (AWS) [3]. The data related to a user is kept in a backend database and the user accesses the social cloud by the access control mechanism [49]. Social network forms a dynamic social cloud by enabling friends to share resources within the context of a social network. Users share their opinions, experiences in various topics and also are able to learn about other users' thoughts and experiences on a particular topic such as health. Social networks are substantial tools for healthcare. It helps patients to receive information and social support, ask advice from other patients with similar disease or medical experts. "e-health" term is used to describe the healthcare services and information delivered through the Internet. The main role of an e-health social network is to find other patients in similar situations and share information about treatments, symptoms and conditions. Some e-health services provide emotional support and some of them provide ability to ask questions to a medical expert [44]. The authors in chapter *Social Media: The Evolution of E-health Services*, analyze the e-health services provided by different Social Media, give an overview of different studies on Social Media in the healthcare sector and a description about the different activities and relationships on Social Media among physicians and patients. They introduce a Hybrid CLoud E-health Services architecture (HCLES) which is capable of providing open, interoperable, scalable, and extensible services for all the e-health activities. Their proposed architecture integrates the use of tele-consulting service of Skype for a direct communication and synchronous data transmission and the cloud platform. The cloud platform consists of Infrastructure as a Service (IaaS) by providing a Web interface that allows patients and physicians accessing virtual machines; Software as a Service (SaaS) by providing all the e-health services offered by Social Media to patients and physicians; Platform as a Service (PaaS) by offering an integrated set of Social Media that provides an online environment for quick communication and collaboration between patients and physicians. The proposed architecture provides the need of supporting existing communities and facilitating their connection by Skype, and creating communities of interests of patients, physicians or hybrid communities that provide/receive emotional and psychological support, medical support, medical information, health care education, and tele-consulting.

5 Conclusion

Recently, SNA has attracted a significant interest in many fields such as sociology, epidemiology, anthropology, social psychology, communication studies, information sciences, etc. This chapter provides definitions of the basic concepts of SNA and briefly introduces the topics in the book. A vast number of topics exist in the SNA field, therefore it is not possible to cover all of them comprehensively. However, this book includes most of the significant works and achieves the following:

- presents background on social networks and SNA.
- reviews the related works and their outcomes obtained on the addressed topics.
- demonstrates various important applications and studies in the areas of social networks, social community mining, social behavior and network analysis.

Through these, the book aims to introduce readers to the area of SNA and to become a reference book for academicians and industrial practitioners.

References

1. Aggarwal CC, Yu PS (2008) An introduction to privacy-preserving data mining. In: Privacy-preserving data mining, vol 34. Springer, Berlin, pp 1–9 [ISBN: 978-0-387-70991-8]
2. Alexa traffic statistics for Facebook (2013) http://www.alexa.com/siteinfo/facebook.com
3. Amazon, (2013) Building facebook applications on aws website. http://aws.amazon.com/solutions/global-solution-providers/facebook/
4. Balkundi P, Harrison DA (2006) Ties, leaders, and time in teams: strong inference about network structures effects on team viability and performance. Acad Manage J 49(1):49–68
5. Barnes JA (1954) Class and committee in a Norwegian island parish. Hum Relat 7:39–58
6. Bonacich P, Lloyd P (2001) Eigenvector-like measures of centrality for asymmetric relations. Soc Netw 23:191–201. doi:10.1016/S0378-8733(01)00038-7
7. Cai D, Shao Z, He X, Yan X, Han J (2005) Community mining from multi-relational networks. In: Jorge A, Torgo L, Brazdil P, Camacho R, Gama J (eds) PKDD. Springer, Berlin, pp 445–452 [ISBN: 3-540-29244-6]
8. Cheek GP, Shehab M (2012) Policy-by-example for online social networks. In: SACMAT. ACM, New York, S. 23–32 [ISBN 978-1-4503-1295-0]
9. Dai BT, Chua FCT, Lim E-P (2012) Structural analysis in multi-relational social networks. In: SDM. SIAM/Omnipress, California, USA, pp 451–462 [ISBN: 978-1-61197-232-0]
10. Fong PWL, Siahaan I (2011) Relationship-based access control policies and their policy languages. In: SACMAT. ACM, New York, S. 51–60 [ISBN 978-1-4503-0688-1]
11. Fortunato S (2010) Community detection in graphs. Phys Rep 486:75–174
12. Freeman LC (1979) Centrality in social networks: conceptual clarification. Soc Netw 1:215–239. doi:10.1016/0378-8733(78)90021-7

13. Motta G, Sfondrini N, Sacco D (2012) Cloud computing: an architectural and technological overview. In: 2012 international joint conference on service sciences (IJCSS). IEEE, New York, pp 23–27
14. Bastian M, Heymann S, Jacomy M (2009) Gephi: an open source software for exploring and manipulating newtorks. In: Proceedings of the AAAI international conference on weblogs and social media (ICWSM'09)
15. Gilks W, Spiegelhalter D (1996) Markov chain Monte Carlo in practice. Chapman & Hall/CRC, London/Boca Raton
16. Girvan M, Newman MEJ (2002) Community structure in social and biological networks. Proc Natl Acad Sci 99:7821–7826
17. Granovetter M (1973) The strength of weak ties. Am J Sociol 78:1360–1380
18. Hagberg AA, Schult DA, Swart PJ (2008) Exploring nework structure, dynamics, and function using networkX. In: Proceedings of the 7th phython in science conference, Pasadena, CA, USA, pp 11–15. http://networkx.lanl.gov/
19. Csardi G, Nepusz T (2006) The igraph software package for complex network research. InterJournal, Complex Systems. http://www.interjournal.org/manuscript_abstract.php?361100992
20. Kozlowski SWJ, Bell BS (2003) Work groups and teams in organizations. In: Borman WC, Ilgen DR, Klimoski R (eds) Handbook of psychology: industrial and organizational psychology, vol 12. Wiley, New York, pp 333–375
21. Kozlowski SWJ, Ilgen DR (2006) Enhancing the effectiveness of work groups and teams. Psychol Sci Public Interest 7(3):77–124
22. Liu K, Terzi E (2008) Towards identity anonymization on graphs. In: Wang JT-L (ed) SIGMOD Conference, pp 93–106. ACM, New York [ISBN: 978-1-60558-102-6]
23. Lovasz L (1993) Random walks on graphs: a survey, Bolyai Soc. Math. Stud. Keszthely, Hungary, 2:1–46
24. Mayo E (1945) The social problems of an industrial civilization. Routledge and Kegan Paul, London
25. Metropolis N, Rosenbluth AW, Rosenbluth MN, Teller AH, Teller E (1953) Equation of state calculations by fast computing machines. J Chem Phy 21:1087–1092. doi:10.1063/1.1699114
26. Milgram S (1967) The small world problem. Psychol Today 22:61–67
27. Moody J, White D (2003) Structural cohesion and embeddedness: a hierarchical concept of social groups. Am Sociol Rev 68(1):103–127
28. Moreno J (1934) Who shall survive? Beacon Press, New York
29. Narayanan A, Shmatikov V (2009) De-anonymizing social neworks. In: IEEE symposium on security and privacy, IEEE Computer Society, pp 173–187
30. Neville J, Jensen D (2000) Iterative classification in relational data. AAAI Press, Texas, USA, pp 13–20
31. Nieminen J (1974) On centrality in a graph. Scand J Psychol 15:322–336
32. O'Madadhain J, Fisher D, White S, Boey Y (2003) The JUNG (java universal Network/Graph) framework. Technical report, UCI-ICS
33. Page L, Brin S, Motwani R, Winograd T (1998) The PageRank citation ranking: bringing order to the web. Stanford Digital Library Technologies Project, California, USA
34. Palla G, Barabasi A-L, Vicsek T (2007) Quantifying social group evolution. Nature 446:664–667. doi:10.1038/nature05670
35. Pfitzmann A, Hansen M (2010) A terminology for talking about privacy by data minimization: anonymity, unlinkability, undetectability, unobservability, pseudonymity, and identity management (version 0.34 vom 10. August 2010)
36. Pons P, Latapy M (2005) Computing communities in large networks using random walks (long version). In: Computer and Information Sciences-ISCIS 2005, pp 284–293
37. Raghavan UN, Albert R, Kumara S (2007) Near linear time algorithm to detect community structures in large-scale networks. Phys Rev E 76:036106
38. Rasti AH, Torkjazi M, Rejaie R, Duffield NG, Willinger W, Stutzbach D (2009) Respondent-driven sampling for characterizing unstructured overlays. In: INFOCOM. IEEE, New York, pp 2701–2705

39. Reagans R, Zuckerman EW (2001) Networks, diversity, and productivity: the social capital of corporate R&D. Teams Organ Sci 12(4):502–517
40. Salganik MJ, Heckathorn DD (2004) Sampling and estimation in hidden populations using respondent-driven sampling. Sociol Methodol 34:193–239. doi:10.2307/3649374
41. Sanna L, Parks C (1997) Group research trends in social and organizational psychology: whatever happened to intragroup research? Psychol Sci 8:261–267
42. Scott J (2005) Social network analysis: a handbook. 2nd edn. Sage Publications, London
43. Stocker K (2013) Battle of the social network climbers. http://www.drapersonline.com/news/ecommerce/webwatch/battle-of-the-social-network-climbers/5052344.article
44. Swan M (2009) Emerging patient-driven health care models: an examination of health social networks, consumer personalized medicine and quantified self-tracking. Int J Environ Res Public Health 6(2):492–525
45. Verykios VS, Bertino E, Fovino IN, Provenza LP, Saygin Y, Theodoridis Y (2004) State-of-the-art in privacy preserving data mining. SIGMOD Rec 33:50–57
46. Warner WL, Lunt PS (1941) The social life of a modern community. Yale University Press, New Haven
47. Wasserman S, Faust K, Iacobucci D, Granovetter M (1994) Social network analysis: methods and applications. Cambridge University Press, Cambridge
48. Watts D, Strogatz S (1998) Collective dynamics of 'small-world' networks. Nature 393(6684):440–442
49. Wooten R, Klink R, Sinek F, Bai Y, Sharma M (2012) Design and implementation of a secure healthcare social cloud system. In: CCGRID. IEEE, New York, pp 805–810 [ISBN: 978-1-4673-1395-7]
50. Ye S, Lang J, Wu F (2010) Crawling online social graphs. In: 2010 12th International Asia-Pacific Web Conference (APWEB), pp 236–242

Ranking Authors on the Web: A Semantic AuthorRank

Lule Ahmedi, Lavdim Halilaj, Gëzim Sejdiu, and Labinot Bajraktari

Abstract Author ranking is growing in popularity since search engines are considering the author's reputation of a Web page when generating search results. A question that naturally arises is whether we should rank authors on the Web as we rank Web pages by considering their links. In addition, over what links to actually calculate author ranking? We have adopted an extended FOAF ontology, the so-called CO-AUTHORONTO ontology, able to represent authors, but also their co-author links on the Web. We further extended CO-AUTHORONTO with PageRank and AuthorRank metrics for ranking authors based on their co-author links. Important to note is that both PageRank and AuthorRank are implemented in Semantic Web Rule Language (SWRL), which represents a novelty and fits well with the semantic modeling of authors and their co-author relationships within FOAF. Preliminary semantic ranking results are demonstrated, showcasing also the huge potential of this ranking approach for adopting it by search engines where our future work will focus.

1 Introduction

Given the "invasion" of social computing applications on the Web, Handler and Berners-Lee envision that we are just at the beginning of this growing and evolving age of "social machines" [1]. Social networks are mainly about linking people [2, 3], and the FOAF ontology has become a core model on the Web for describing

L. Ahmedi (✉) • L. Halilaj • G. Sejdiu • L. Bajraktari
Faculty of Electrical and Computer Engineering, University of Prishtina, Prishtina, Republic of Kosovo
e-mail: lule.ahmedi@uni-pr.edu; lavdim.halilaj@uni-pr.edu; gezim.sejdiu@uni-pr.edu; labinot.bajraktari@uni-pr.edu

Ş. Gündüz-Öğüdücü and A.Ş. Etaner-Uyar (eds.), *Social Networks: Analysis and Case Studies*, Lecture Notes in Social Networks, DOI 10.1007/978-3-7091-1797-2_2,

the profile data of people and their social links [4]. The focus on the Web has now shifted from the hypertext links (aka `href`) to people links (aka `knows`). Search engines are also shifting from ranking only pages, to ranking pages improved by considering the ranking of people authoring those pages. As Google and Bing announced, the reputation of authors of Web pages will influence the tops displayed in search results.

A question that naturally arises is whether we should rank authors on the Web as we rank Web pages by considering their links. In addition, over what links to actually calculate author ranking? We have adopted an extended FOAF ontology, the so-called CO-AUTHORONTO ontology [5], able to represent authors, but also their co-author links on the Web. It is turning into common practice to accompany publishing documents on the Web with publishing their co-authorship data as well on the Web. We therefore concentrate here in links connecting authors, namely co-author links, and propose a model which extends CO-AUTHORONTO with PageRank [6] and AuthorRank [7] metrics for ranking authors to gauge their reputation based on their co-author links. Both PageRank and AuthorRank are implemented in SWRL [8, 9]. According to [7], AuthorRank is an advanced version of PageRank which takes into account the weight of co-author links when ranking, i.e., it does not assume that all links have the same weight.

Referring to [1], the growing Semantic Web provides necessary support for social computing technologies. Our framework exactly utilizes the Semantic Web technologies to support ranking on the Web of authors, like in online co-authorship networks (DBLP, ACM, IEEE), or in social networks (FOAF), or in general when searching through search engines.

The rest of the paper is organized as follows: We first introduce related work, followed by the background knowledge required to understand the rest of the paper, where CO-AUTHORONTO, our earlier extension of FOAF to capture the semantics of weighted co-authorship networks is also covered. Further we detail our contribution: the extension of FOAF, namely of CO-AUTHORONTO, and rule definitions in SWRL, to rank authors by PageRank and AuthorRank. Then an evaluation of our ranking framework is presented. Finally we conclude by highlighting the main contribution of this work, and the future work to follow.

2 Related Work

In [10], a study to rank people based on FOAF data (referred to as AuthorRank there), and combining people ranking with co-citation analysis is presented. Ranking has been carried out by storing and querying in Yars [11], a Semantic Web platform. Another framework [12] deploys Semantic Web for the representation and people ranking by centrality measures of social network analysis (SNA) expressed in form of SPARQL queries. Applying SNA metrics to the co-authorship networks on the Web has been investigated in [13, 14] in finding interesting facts about authors of scientific publications and relationships among them. Liu et al. [7] have introduced ranking of co-authorship networks including their ranking by PageRank and AuthorRank.

To the best of our knowledge, there is no approach to date which is based solely on the Semantic Web to rank people in social networks or co-authorship networks based on PageRank and AuthorRank. Nor is there an approach of considering (SWRL) rules of Semantic Web to express and reason over PageRank and AuthorRank for ranking people on the Web.

3 Background

Next, a set of preliminary knowledge required for the rest of the work is briefly introduced:

- Definition of ranking measures like PageRank and AuthorRank for authors in co-authorship networks.
- Modeling co-authorship networks on the Web using CO-AUTHORONTO, an extended FOAF ontology.
- SNA metrics like exclusivity, frequency, or weight, and their implementation for co-authorship networks.

3.1 Author Ranking via PageRank and AuthorRank

Let us first recall the definitions according to [7] for PageRank and its weighted version, AuthorRank, of the directed weighted co-authorship graph model we adopt to rank authors in co-authorship networks on the Web. Let $A = \{a_1,.., a_k,.., a_n\}$ denote the set of n authors.

Definition 1 (PageRank) The PageRank of an author a_i is given as follows:

$$PR(a_i) = (1-d)/n + d\sum_{j=1}^{n}\left(PR(a_j)\cdot\frac{1}{C(j)}\right)$$

Definition 2 (AuthorRank) The AuthorRank of an author a_i is given as follows:

$$AR(a_i) = (1-d)/n + d\sum_{j=1}^{n}\left(AR(a_j)\cdot w_{j,i}\right)$$

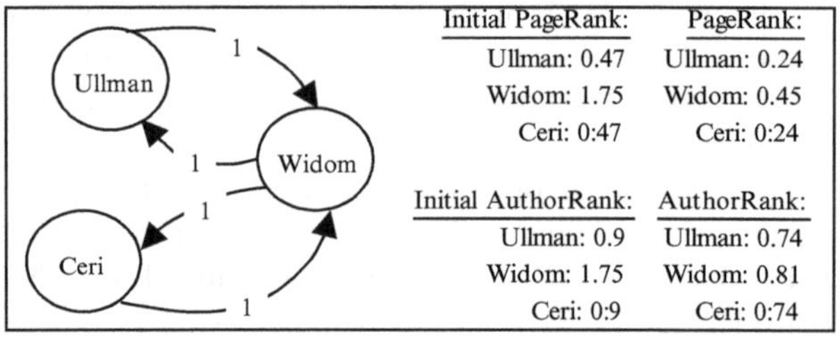

Fig. 1 The weighted directed co-authorship network of *Example 1* and ranking values

where $AR(a_j)$ corresponds to AuthorRank of the backlinking co-author node, and $w_{i,j}$ to the edge weight between co-authors a_j and a_i. Analogous to the initial PageRank, the first initial iteration of the AuthorRank algorithm assumes $AR(a_j) = 1$ for $j = 1 \ldots n$.

One may think of AuthorRank as a generalization of PageRank by substituting $1/C(j)$ with $w_{j,i}$. According to [7], AuthorRank better reveals status of actors than centrality SNA measures and PageRank.

Example 1 The running example we will use throughout this paper consists of two publications and their respective authors as follows:

Publication	Authors		
First Course in Database Systems	Ullman	Widom	
Active Database Systems		Widom	Ceri

Figure 1 provides the metrics calculated following the definitions 1 and 2 for the example above.

3.2 CO-AUTHORONTO: *A FOAF Extension to Model Co-authorship Networks on the Web*

In our earlier work [5], we extended FOAF into CO-AUTHORONTO (Fig. 2) to model rich co-author relations with attributes such as *weight*, as well as other knowledge relevant to collaboration schemes in the scientific community, as suggested in [7].In addition to FOAF, the CO-AUTHORONTO integrates the AKT Reference Ontology [15], its Portal and Support sub-ontologies describing people, publications, etc.

Two main classes in the CO-AUTHORONTO ontology are `Person` and `Publication`, with their respective subclasses, such as `Researcher`, or `Student` for the former, or, `Serial-Publication`, or `Book` for the later, all inherited either from the AKT ontologies, or FOAF.

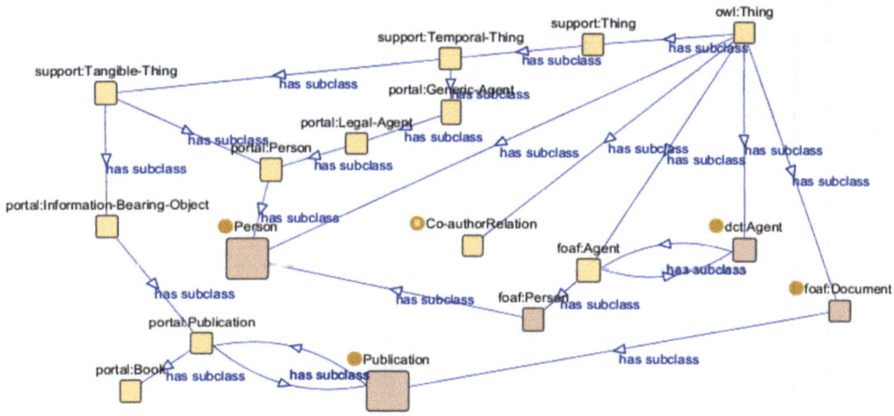

Fig. 2 An excerpt of the CO-AUTHORONTO ontology—how it relates to the `foaf`, `portal` and `support` (class hierarchy) ontologies

Each `Publication` instance may have (multiple) values of type `Person` on the `hasAuthor` property. The `hasExclusivity` property stores the exclusivity degree on a given publication, as explained in Table 1. All co-author pairs of the same publication should share a common value for the exclusivity. The property `nrAuthors`, as its naming reveals, is designed to record data about the number of authors of a given publication. For `hasExclusivity` and `nrAuthors`, their values are inferred once their corresponding rule implementations in SWRL (Table 1) are evaluated.

Figure 3 illustrates the use of OWL n-ary relations [16] for representing the `hasCo_author` property of the `Person` class. It links an individual `Person` to another individual `Person`, as well as to another value, *weight*, which characterizes that `hasCo-author` relation.

3.3 *SNA Metrics on Co-authorship Networks on the Web*

Let again $A = \{a_1,.., a_k,.., a_n\}$ denote the set of n authors, whereas $\mathrm{P} = \{p,.., p_k,.., p_m\}$ the set of m publications. Let $f(p_k)$ define the number of authors of publication p_k. Then in Table 1, definitions of SNA metrics, i.e., the exclusivity, co-authorship frequency, total frequency, and weight [7], as well as their respective rule definitions in SWRL according to our earlier work [5] are summarized. Once these metrics' rules are evaluated by an inference engine, their inferred values are assigned to the CO-AUTHORONTO ontology.

Table 1 SNA metrics and their rule definitions in Co-AuthorOnto

SNA metric		SWRL rule expression
Co-author relation	Not applicable	`Publication(?p) ∧ hasAuthor(?p,?a1) ∧ abox:hasURI(?a1,?a1URI) ∧ swrlb:substringAfter(?a1name, ?a1URI,"#") ∧ hasAuthor(?p,?a2) ∧ abox:hasURI(?a2,?a2URI) ∧ swrlb:substringAfter(?a2name,?a2URI,"#") ∧ swrlb:notEqual(?a1name,?a2name) ∧ swrlx:makeOWLThing(?rel,?a1,?a2) → Co-authorRelation(?rel) ∧ hasCo-author(?a1,?rel) ∧ hasCo-authorValue(?rel,?a2)`
Number of authors of a given publication	$f(p_k)$	`Publication(?p) ∧ hasAuthor(?p,?a) ° sqwrl:makeSet(?s,?a) ∧ sqwrl:groupBy(?s,?p) ° sqwrl:size(?size,?s) → nrAthors(?p,?size)`
Exclusivity per publication: Degree to which authors a_i, a_j have an exclusive co-authorship relation for a particular p_k publication. This definition gives more weight to co-author relationships in publications with fewer total co-authors than publications with large numbers of co-authors, i.e., it weighs the co-author relation in terms of how exclusive it is	$g_{i,j,k} = \frac{1}{(f(p_k)-1)}$	`Publication(?p) ∧ nrAuthors(?p,?nr) ∧ swrlb:greaterThanOrEqual(?nr,2) ∧swrlm:eval(?e,"1/(nr-1)",?nr) →hasExclusivity(?p,?e)`

Co-authorship frequency: Sums up the exclusivity values $g_{i,j,k}$ for that same pair i,j of authors across all publications k (k = 1 . . . m) where they appear as co-authors a_i, a_j. This gives more weight to authors who co-publish more publications together, and do so exclusively	$c_{ij} = \sum_{k=1}^{m} g_{i,j,k}$	Publication(?p) ∧ nrAuthors(?p,?nr) ∧ swrlb:greaterThanOrEqual(?nr, 2) ∧ hasAuthor(?p,?a1) ∧ hasAuthor(?p,?a2) ∧ hasCo-author(?a1,?rel) ∧ hasCo-authorValue(?rel,?a2) ∧ abox:hasURI(?a1,?a1URI) ∧ abox:hasURI(?a2,?a2URI) ∧ swrlb:notEqual(?a1URI,?a2URI) ∧ sameAs(?a1,?a1) ∧ sameAs(?a2,?a2) ∧ hasExclusivity(?p,?e) ° sqwrl:makeSet(?s,?e) ∧ sqwrl:groupBy(?s,?a1,?a2) ° sqwrl:sum(?f,?s) → hasCo-authorFrequency(?rel,?f)
Total co-authorship frequency: The sum of all co-authorship frequency values c_{ik} over a particular author ai and all of its co-authors a_k (k = 1 . . . n) in whatever publications where a_i appears as author	$c_i = \sum_{k=1}^{n} c_{ik}$	Person(?a1) ∧ hasCo-author(?a1,?rel) ∧ hasCo-authorValue(?rel,?a2) ∧ hasCo-authorFrequency(?rel,?f) ° sqwrl:makeBag(?s,?f) ∧ sqwrl:groupBy(?s,?a1) ° sqwrl:sum(?totalFa1,?s) → hasTotalFrequency(?a1,?totalFa1)
Co-author weight: Normalization by taking into account the total co-authorship frequency c_i of a given author a_i when calculating the co-authorship frequency c_{ij} between that author a_i	$w_{ij} = \frac{c_{ij}}{c_i}$	hasCo-author(?a1,?rel) ∧ hasCo-authorValue(?rel,?a2) ∧ hasTotalFrequency(?a1,?totalFa1) ∧ hasCo-authorFrequency(?rel,?a12f) ∧ swrlm:eval(?w,"a12f/totalFa1",?a12f,?totalFa1) → hasCo-authorWeight(?rel,?w)

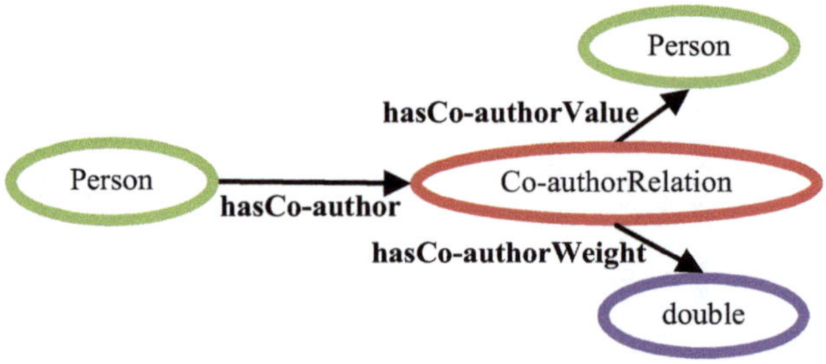

Fig. 3 N-ary `hasCo-author` property in Co-AuthorOnto (class view)

4 Rank Authors in Web Co-authorship Networks

Now we describe in detail our proposal to further extend FOAF, namely Co-AuthorOnto [5] which is originally built on top of FOAF and AKT ontologies and supports SNA metrics [7] just introduced in the previous section, to support PageRank and AuthorRank for ranking authors,. The main novelty of our proposal, i.e. the Semantic Web rule expressions for computing PageRank and AuthorRank are also introduced.

4.1 FOAF Extended with PageRank and AuthorRank

Let us next introduce our proposal to extend the Co-AuthorOnto (Fig. 1) ontology [5] with new ontological constructs which enable ranking by PageRank and AuthorRank of authors in directed weighted co-authorship networks.

The `Person` class is among extended concepts to support ranking, and its extension is defined as following:

```
Person   a  owl:Class;
 rdfs:subClassOf
   [a  owl:Restriction;
       owl:allValuesFrom: float;
       owl:onProperty: hasPR].
   rdfs:subClassOf
 [a  owl:Restriction;
       owl:allValuesFrom: float;
       owl:onProperty: hasAR];
   rdfs:subClassOf
   [a  owl:Restriction;
       owl:allValuesFrom: int;
       owl:onProperty:
      hasTotalNrCo-Authors].
```

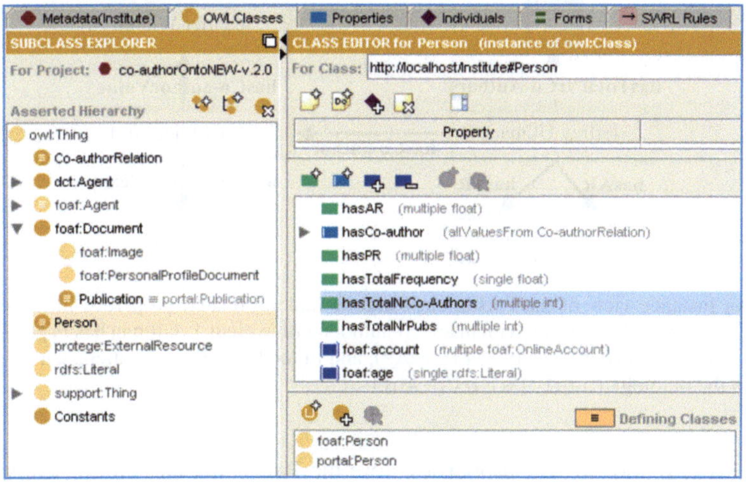

Fig. 4 Some properties of the `Person` class introduced in our ontology

- Added new datatype properties, `hasPR` and `hasAR`, both of domain `Person`, and their range set to `float`.
- Added a new datatype property `hasTotalNrCo-Authors`, of domain `Person` and the `int` range.

Further, a new class `Constants` is introduced:

```
:Constants   a  owl:Class;
        rdfs:subClassOf
        [a  owl:Restriction;
            owl:allValuesFrom: int;
         owl:onProperty:totalNrAuthors].
```

- Defined the datatype property `totalNrAuthors`, of domain `Constants`, and its range set to `int`.
- Instantiated a new single `Constants` individual, the `Constant1` individual, to hold the sole expected value for the `totalNrAuthors` property in our ontology (aka assigning a value to a global variable in programming languages).

Notice in Fig. 4 (a snapshot of our ontology in Protégé [17], an open-source Semantic Web system in Java we used for modeling and development) the class `Person` in the class hierarchy on the left window, its definition as equivalent class to classes `portal:Person` of AKT ontology, and `foaf:Person` of FOAF ontology shown in the right bottom window, and a list of its properties and restrictions on them shown in the right top window.

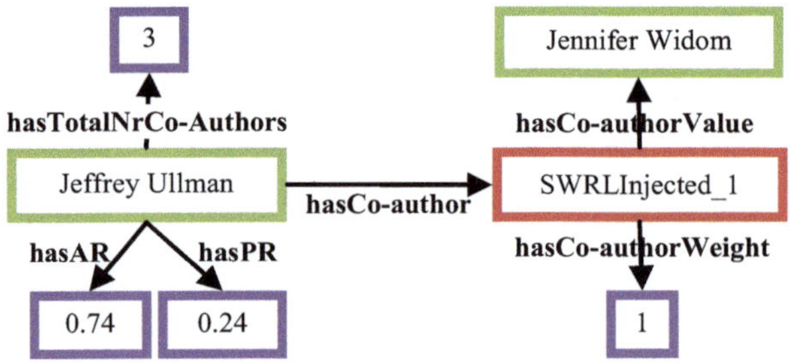

Fig. 5 An instance view of the Person class and its ranking-related properties in our ontology: *Example 1* excerpt: Ullman and Widom are co-authors with weight 1; Ullman has PageRank value 0.24 (hasPR), AuthorRank value 0.74 (hasAR), and in total shares publications with three other authors in the ontology (hasTotalNrCo-Authors)

The hasPR and hasAR properties store the PageRank and AuthorRank values, respectively, for a given author of type Person. The hasTotalNrCo-Authors holds the total number of authors acting as co-authors for a given author in whatever publication in the ontology. The property totalNrAuthors, as its naming reveals, is designed to record data about the number of (distinct) authors in total for the whole ontology. For all these four properties, hasPR, hasAR, hasTotalNrCo-Authors, and totalNrAuthors, their values are inferred by reasoning through SWRL rules over the ontology, as will be introduced in the next section.

Figure 5 illustrates an author, Ullman, and its values for the ranking-related properties. The SWRLInjected_1 acts somewhat as an abstract relation instance: It serves merely to link from the source instance (Ullman) to this relation instance (SWRLInjected_1), and from this relation instance (SWRLInjected_1), to the target instance (Widom).

4.2 *SWRL Rules to Calculate PageRank and AuthorRank*

Moving from ontologies upwards the Semantic Web layer architecture, we will encounter the logic layer with rules which may capture semantics that cannot be directly captured using description logic of OWL [18].

We have defined rules in SWRL to implement PageRank and AuthorRank metrics [7] in conformance with the related ontological constructs we defined in the extended Co-AuthorOnto ontology. Next we detail each of these rules and the metrics of the weighted co-authorship graph model they implement.

The values of these metrics, once pre-calculated through an inference engine, are assigned to the existing ontology annotations.

Rule 1 (Total Number of Authors) This rule will find the overall number of `Person` individuals in the ontology.

```
portal:Person(?a) ° sqwrl:makeSet(?s,?a) °
sqwrl:size(?n,?s) → totalNrPersons(Constant1,?n)
```

It will collect all persons `?a`, i.e., authors of all existing publications, into a common set `?s` (the `makeSet` built-in predicate), and then extract the total number of elements, `?n`, in the set (the `size` built-in predicate). The number obtained will then be asserted to the single existing `totalNrPersons` property of the `Constant1` individual in the ontology.

Rule 2 (Total Number of Co-authors of a Given Author) Next, the rule in SWRL to calculate the number of co-authors for each given person is presented.

```
portal:Person(?a2) ∧ hasCo-author(?a2,?rel) ∧ hasCo-
authorValue(?rel,?a1) ∧ abox:hasURI(?a1,?a1URI) ∧
abox:hasURI(?a2,?a2URI) ∧ swrlb:notEqual(?a1URI,?a2URI) ∧
sameAs(?a1,?a1) ∧ sameAs(?a2,?a2) ° sqwrl:makeBag(?s,?a1) ∧
qwrl:groupBy(?s,?a2) ° sqwrl:size(?allca,?s) → hasTotal-
NrCo-Authors(?a2,?allca)
```

Analogous to Rule 1, the collection built-ins `makeBag` (bags allow duplicate elements, sets do not) and `groupBy` are applied, but this time targeted to collect co-author pairs, group them per author (person), and finally calculate the number of elements (the `size` built-in predicate), i.e., of co-authors' that belong to that given group. The result will assert a value on the `hasTotalNrCo-Authors` property of the given author individual,`?a1`, once per each such author (i.e., per group created). The `notEqual` built-in predicate does exclude a combination of an author with himself (`a1,a1`) as a possible co-author pair.

Rule 3 (Initial PageRank) Following Definition 1, a rule to calculate the first iteration of the PageRank over authors, i.e., over a given author `?ai`, is shown below, and proceeds as follows:

1. It considers all backlinking co-author relations to `?a1`, e.g., that `?rel` connecting author `?aj` to author `?ai`.
2. Applies the sum built-in predicate to sum up into `?ally` the contributions of all backlinking co-authors to `?ai`, e.g., the contribution `?yj=1/?cj` of co-author `?aj`, where `?cj` is the number of co-author relations going out of co-author `?aj`.
3. Evaluates the PageRank formulae according to Definition 1 taking into account also the dumping factor which is set to 0.85, as well as the total number of authors in the ontology as provided by the `totalNrPersons` predicate (cf. Rule 2).

```
hasCo-author(?aj,?rel) ∧ hasCo-authorValue(?rel,?ai) ∧
abox:hasURI(?aj,?pjURI) ∧ abox:hasURI(?ai,?piURI) ∧
swrlb:notEqual(?ajURI,?aiURI) ∧ sameAs(?aj,?aj) ∧
sameAs(?ai,?ai) ∧hasTotalNrCo-Authors(?aj,?cj) ∧ totalNr-
Persons(Constant1,?n) ∧ swrlm:eval(?yj,"1/cj",?cj)
˚sqwrl:makeBag(?s,?yj) ∧ sqwrl:groupBy(?s,?ai) ˚
sqwrl:sum(?ally,?s) ∧ swrlm:eval(?pri,"(0.15 / n) + 0.85 *
ally",?n,?ally) → hasPR(?ai,?pri)
```

The `eval` is just another built-in of the `swrlm` library of SWRL to evaluate the given mathematical formula.

Rule 4 (PageRank) Once the initial PageRank SWRL rule as defined above is evaluated, the system further evaluates the actual PageRank rule for each (other) author recursively. The only difference of this rule to the initial PageRank rule (Rule 3) is in step 2:

- The contributions of all backlinking co-authors to the author under consideration, e.g., to author `?ai`, involve also the actual PageRank values of backlinking co-authors, e.g., the PageRank of co-author `?aj`. In other words, the contribution `?yj=prj/?cj`, where `prj` is the actual PageRank value of that co-author `?aj`, and `?cj` is the nr. of co-author relations going out of co-author `?aj`. At the first iteration of this rule, the PageRank values at the body of the rule are those inferred by the initial PageRank rule.

```
hasPR(?aj,?prj) ∧ hasCo-author(?aj,?rel) ∧  hasCo-
authorValue(?rel,?ai) ∧  abox:hasURI(?aj,?ajURI) ∧
abox:hasURI(?ai,?aiURI) ∧  swrlb:notEqual(?ajURI,?aiURI) ∧
sameAs(?aj,?aj) ∧  sameAs(?ai,?ai) ∧hasTotalNrCo-
Authors(?aj,?cj) ∧totalNrPersons(Constant1,?n) ∧
swrlm:eval(?yj,"prj/cj",?prj,?cj) ˚ sqwrl:makeBag(?s,?yj)
∧ sqwrl:groupBy(?s,?ai) ˚ sqwrl:sum(?ally, ?s) ∧
swrlm:eval(?pri,"0.15 / n + 0.85 * ally",?n,?ally) →
hasPR(?ai,?pri)
```

This rule will iterate once over all authors and their co-author pairs found in the ontology (the `?rel` individuals of the `Co-authorRelation` class), compute their actual PageRank values, and accordingly augment them using the `hasPR` properties which represent PageRank.

Rule 5 (Initial AuthorRank) The SWRL rule which implements the initial AuthorRank is distinct from the SWRL rule which implements the initial PageRank in step 2 (cf. Rule 3), namely in:

- The contributions which it sums up into `?ally` of all backlinking co-authors to `?ai`. For instance, the contribution of co-author `?aj` is `?yj=?wji`, where

?wji is the weight of the co-author relation between the author ?ai under consideration and its current co-author ?aj.

```
hasCo-author(?aj,?rel) ∧ hasCo-authorValue(?rel,?ai) ∧ has-
Co-authorWeight(?rel,?wji) ∧ abox:hasURI(?aj,?pjURI) ∧
abox:hasURI(?ai,?piURI) ∧ swrlb:notEqual(?ajURI,?aiURI) ∧
sameAs(?aj,?aj) ∧ sameAs(?ai,?ai) ∧ sqwrl:makeBag(?s,?wji)
∧ sqwrl:groupBy(?s,?ai) ° sqwrl:sum(?sumj,?s) ∧ totalNrPer-
sons(Constant1,?n) ∧ swrlm:eval(?ari,"(0.15 / n) + 0.85 *
sumj",?n,?sumj) → hasAR(?ai,?ari)
```

This corresponds to exactly substituting what was 1/C(j) in initial PageRank with $w_{j,i}$ in initial AuthorRank.

Rule 6 (AuthorRank) Finally, the following rule calculates the AuthorRank (Definition 2) of every author in the ontology given: (1) the co-authorship network, and (2) the initial AuthorRank values of its co-authors calculated by Rule 5. It reasons over values assigned to the hasCo-authorWeight property of a particular ?rel co-author relationship between ?aj and ?ai, uses the makeBag built-in predicate to collect them into a bag rather than a set (bags allow duplicate values, whereas sets do not), groups the bag values by author ?ai, and calculates the sum of values on each group separately, yielding thus the value for AuthorRank of the respective author (bound to that respective author on his hasAR property in our ontology).

```
hasCo-author(?aj,?rel) ∧ hasCo-authorValue(?rel,?ai) ∧ has-
Co-authorWeight(?rel,?wji) ∧ abox:hasURI(?aj,?pjURI) ∧
abox:hasURI(?ai,?piURI) ∧ swrlb:notEqual(?ajURI,?aiURI) ∧
sameAs(?aj,?aj) ∧ sameAs(?ai,?ai) ∧ hasPR(?aj,?arj) ∧
swrlm:eval(?yj, "arj*wji", ?arj,?wji) °
sqwrl:makeBag(?s,?yj) ∧ sqwrl:groupBy(?s,?ai) °
sqwrl:sum(?sumj,?s) ∧ totalNrPersons(Constant1, ?n) ∧
swrlm:eval(?ari,"(0.15)/n + 0.85 * sumj",?n,?sumj) →
hasAR(?ai,?ari)
```

As one may notice, the only difference of this rule to the initial AuthorRank rule (cf. Rule 5) is following:

- The contributions of all backlinking co-authors to the author under consideration, e.g., to author ?ai, involve also the actual AuthorRank values of backlinking co-authors, e.g., the AuthorRank of co-author ?aj. In other words, the contribution ?yj=arj*wji, where arj is the actual AuthorRank value of that co-author ?aj, and ?wji is the weight of the co-authorship relation between co-authors ?aj and ?ai. At the first iteration of this rule, the AuthorRank values at the body of the rule are those inferred by the initial AuthorRank rule.

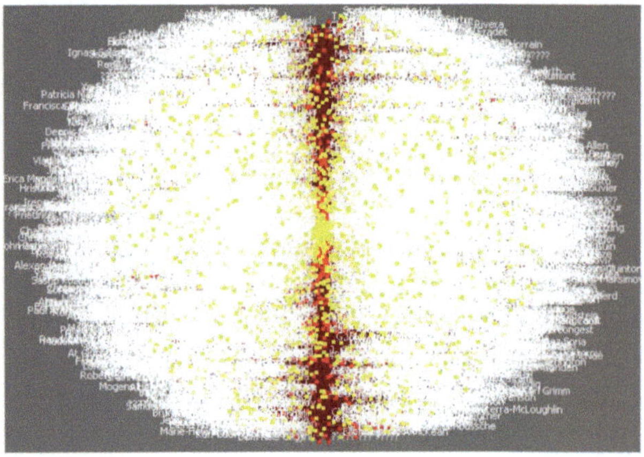

Fig. 6 Co-authorship network of the SEEU library: authors (in *red*), co-authors (in *yellow*), and their links (in *white*)

5 Evaluation Results

The analyses run by ORA [19], a SNAtool, demonstrates there is a dense 3D cloud of 12,482 nodes and 10,984 co-author edges in our test-bed of real-world data (Fig. 6) extracted from the library of the South East European University (SEEU). Evaluation results on these data will be introduced later in Sect. 5.2.

5.1 Evaluation Against a Simple Dataset

Our approach has first been tested on the small sample example, referred to as the running example (Example 1) in this work. The calculation within each iteration (five iterations as sufficient for illustration) of PageRank values according to the Definition 1 formulae (see Sect. 3.1) is detailed in Table 2.

Note that for each author a_i in Table 2, columns $PR(a_{j1})$ and C(j1), and $PR(a_{j2})$ and C(j2) represent the corresponding jth members, j1 and j2, of the sum in Definition 1:

$$sum = \sum_{j \in (j1, j2)} \left(PR(a_j) \cdot \frac{1}{C(j)} \right)$$

For instance, in case of a_1: Ullman as author (see 1st row of iteration 1 in Table 2), the sum evaluates to:

$$sum = PR(a_2) \cdot \frac{1}{C(2)} + PR(a_3) \cdot \frac{1}{C(3)} = \frac{1.00}{2} + 0 = 0.50$$

Table 2 Calculation in five iterations of PageRank over co-author network of *Example 1*

	Author a_i	d	n	$PR(a_{j1})$	C(j1)	$PR(a_{j2})$	C(j2)	sum	$PR(a_i)$
Iteration 1	a_1: Ullman	0.85	3	1.00	2	No link		0.50	0.48
	a_2: Widom	0.85	3	1.00	1	1.00	1	2.00	1.75
	a_3: Ceri	0.85	3	1.00	2	No link		0.50	0.48
Iteration 2	a_1: Ullman	0.85	3	1.75	2	No link		0.88	0.79
	a_2: Widom	0.85	3	0.48	1	0.48	1	0.95	0.86
	a_3: Ceri	0.85	3	1.75	2	No link		0.88	0.79
Iteration 3	a_1: Ullman	0.85	3	0.86	2	No link		0.43	0.41
	a_2: Widom	0.85	3	0.79	1	0.79	1	1.59	1.40
	a_3: Ceri	0.85	3	0.86	2	No link		0.43	0.41
Iteration 4	a_1: Ullman	0.85	3	1.40	2	No link		0.70	0.64
	a_2: Widom	0.85	3	0.41	1	0.41	1	0.83	0.75
	a_3: Ceri	0.85	3	1.40	2	No link		0.70	0.64
Iteration 5	a_1: Ullman	0.85	3	0.75	2	No link		0.38	**0.37**
	a_2: Widom	0.85	3	0.64	1	0.64	1	1.29	**0.14**
	a_3: Ceri	0.85	3	0.75	2	No link		0.38	**0.37**

where $PR(a_3) \cdot \frac{1}{C(3)}$ equals to 0 since there is no co-author link of author a_1 to author a_3: the only co-author to a_1 is author a_2. Further, the PageRank value 0.48 of author a_1 at this iteration is obtained as:

$$PR(a_1) = \frac{(1-d)}{n} + d \cdot sum = \frac{(1-0.85)}{3} + 0.85 \cdot 0.50 = 0.48$$

In a similar way, for author a_2: Widom, the sum is calculated (see Table 2, 2nd row of iteration 1) as follows:

$$sum = PR(a_1) \cdot \frac{1}{C(1)} + PR(a_3) \cdot \frac{1}{C(3)} = \frac{1.00}{1} + \frac{1.00}{1} = 2.00$$

where it is evident that both of its co-authors a_1 and a_3 contribute to the end result. Finally, the PageRank value $PR(a_2)$ of author $a_2$yields the value 1.75.

The rest of PageRank values shown in Table 2 are analogously calculated for all three authors, a_1, a_2, and a_3, and along all iterations.

We evaluated our semantic PageRank rules, Rules 3 and 4, against PageRank values shown in Table 2 and over the same data set (cf. Example 1), and it proved that our semantic PageRank framework performs correctly: compare the results of our semantic PageRank—its screenshot in Fig. 7, with the last iteration (5th iteration) results provided in Table 2 in bold.

Query 1 (PageRank) The expression displayed within the Rule field of Fig. 7: is a Semantic Query-Enhanced Web Rule Language (SQWRL) query used to retrieve

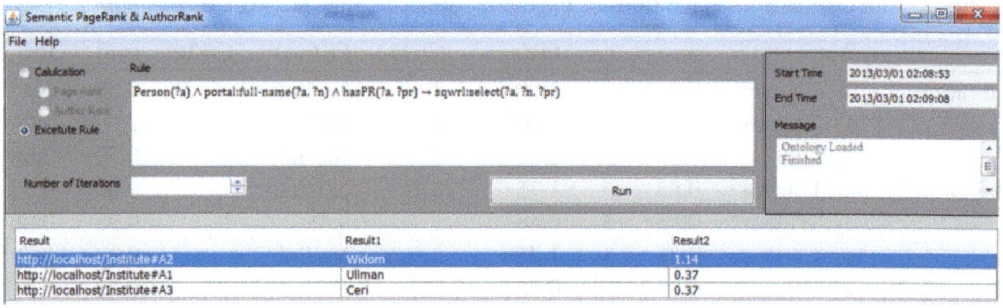

Fig. 7 Ranking according to our semantic PageRank over publications and their authors dataset of *Example 1*

```
Person(?a) ∧ portal:full-name(?a, ?n) ∧ hasPR(?a, ?pr) →
sqwrl:select(?a, ?n, ?pr) ∧ sqwrl:orderByDescending(?pr)
```

PageRank values of authors of Example 1which are calculated by Rules 3 and 4 earlier in our system, and sort them in the descending order by PageRank. The result of this query (Fig. 7) are three rows of authors (column Result lists URIs of authors; column Result1 lists names of authors), and their PageRank values (column Result2). According to PageRank values, author Widom is ranked 1st—its PageRank is 1.14, whereas authors Ullman and Ceri are ranked in the 2nd place—they share the same value 0.3.

5.2 *Real-World Preliminary Evaluation Results*

Next we show the evaluation of our ranking SWRL rules over a real-world co-authorship network. Experiments have been conducted over the co-authorship data we generated from the library database of the South East European University (SEEU) which stores the bibliography of books and other publications available for loan from the university. It contains 12,774 `Publication` individuals, and 11,607 `Person` individuals acting as authors of publications. The experiment was run on a 2 × Quad Core 2.5 GHz machine with 16 GB RAM and Windows Server 2003 2R installed. Centrality measures generated with ORA of the same co-authorship network are given in Table 3.

The SWRL rules in our framework did infer that there are 12,732 co-author relationships (the total number of `hasCo-authorValue` properties in our ontology) as a special type of the FOAF `knows` relationship. Next we present results of our benchmarks on ranking metrics, PageRank and AuthorRank, calculated in our framework of SWRL rules over the extended CO-AUTHORONTO representation of the SEEU library.

Table 3 Top ten ranked authors by their centrality measures

Number	Betweenness centrality	Closeness centrality	Degree centrality
1	Baki Ymeri	William Shakespeare	Ilia Kristo
2	Ralph Mowat	Bardhyl Ceku	William Shakespeare
3	William Shakespeare	Arjan Abazi	Theo Scherling
4	David Cotton	Artan Duka	Miroslav Hadzic
5	Kenneth E Clow	Mustafa Ibrahimi	Clare West
6	Ali Aliu	Clare West	Rebecca Hughes
7	Anne Frank	Miso Nikolov	Ismail Kadare
8	Richard M Hodgetts	Gadaf Rexhepi	Emily Dickinson
9	Gabriel Almond	Ralph Mowat	Erion Kristo
10	Penelope Lively	Theo Scherling	Diane Mowat

Table 4 Top ten ranked authors by four distinct ranking measures. Same ranked authors are highlighted in bold

Number	Author name	PageRank	Author name	AuthorRank
1	**William Shakespeare**	5.513	**William Shakespeare**	5.835
2	**Clare West**	4.131	**Clare West**	4.712
3	Diane Mowat	3.610	**Ismail Kadare**	3.899
4	Ali Aliu	3.214	**Diane Mowat**	3.802
5	Ismail Kadare	3.065	**Ali Aliu**	3.652
6	Honoré de Balzac	2.945	Honoré de Balzac	3.382
7	Erion Kristo	2.943	Erion Kristo	3.283
8	John Escott	2.900	John Escott	3.186
9	Arbër Çeliku	2.891	Noam Chomsky	3.033
10	Rex Gibson	2.799	Arbër Çeliku	2.998
Number	Author name	Nr co-authors	Author name	Total co-authorship frequency
1	**William Shakespeare**	24	**William Shakespeare**	18.5
2	**Clare West**	19	**Clare West**	18
3	**Ismail Kadare**	17	**Ismail Kadare**	14.5
4	**Diane Mowat**	14	**Diane Mowat**	12
5	Noam Chomsky	13	**Ali Aliu**	12
6	Honoré de Balzac	12	John Escott	11
7	Ali Aliu	12	Honoré de Balzac	10
8	Erion Kristo	12	Rex Gibson	9.5
9	John Escott	12	Erion Kristo	9
10	Rex Gibson	11	Noam Chomsky	8

PageRank over SEEU Library The same SQWRL query expression as defined in Query 1 is used to find the computed PageRank values of authors of publications in the SEEU library.

The result of this query is a table of cardinality 7,006. The top ten authors on their PageRank values are shown in Table 4.

Query 2 (AuthorRank) An SQWRL query expression similar to Query 1 is used to check AuthorRank values of authors in the SEEU library which have already been computed by Rules 5 and 6 we introduced in the previous section.

```
Person(?a) ∧ portal:full-name(?a,?n) ∧hasAR(?ar)
→ sqwrl:select(?a, ?n, ?ar)∧ sqwrl:orderByDescending(?ar)
```

The top ten authors on their AuthorRank values are provided in Table 4 next to their PageRank values.

Comparing PageRank and AuthorRank In order to compare PageRank and AuthorRank values of authors in the SEEU library to each other, and how two other co-author metrics, i.e. total co-authorship frequency and total number of co-authors of each given author influence their corresponding PageRank and AuthorRank values, we pose additional two SQWRL queries over SEEU library data as follows:

```
Person(?a) ∧ portal:full-name(?a, ?n) ∧
hasTotalNrCo-Authors(?a,?tnc) →
sqwrl:columnNames("author","name","nr co-authors") ∧
sqwrl:select(?a, ?n,?tnc) ∧ sqwrl:orderByDescending(?tnc)∧
Person(?a) ∧ portal:full-name(?a, ?n) ∧
hasTotalFrequency(?a,?tf) →
sqwrl:columnNames("author","name","total co-authorship fre-
quency") ∧ sqwrl:select(?a, ?n,?tf) ∧
sqwrl:orderByDescending(?tf)
```

As one may also reason from the list in Table 4, there are cases when an author, e.g. Ismail Kadare, has more co-authors (might share diverse roles, like writer, editor, etc.), as well as more exclusive co-author relations (the higher total co-authorship frequency), but has lower PageRank and is hence ranked lower in our table (position 4) than is, e.g. Diane Mowat (position 3) with a higher score for PageRank. Thus, a sum of less but more powerful co-author relations (originating from powerful co-authors) may outperform a higher in number but weaker in power co-author relations.

In contrary, different from PageRank, ranking by AuthorRank is sensitive to the weight of co-author relations, and hence more close to ranking by the total co-authorship frequency in terms that the later (total co-authorship frequency)is sensitive to the exclusivity of its co-author relations (similar to weight). Observe in Table 4 in bold that five top ranked authors are exactly the same ones for both AuthorRank and total co-authorship frequency ranking algorithms.

What AuthorRank yet differentiates from the total co-authorship frequency is that it considers ranking of its co-authors taking into consideration also the weights of their co-authors to their further co-authors, whether the former measure (total co-authorship frequency) relies on the weight of its direct co-author relations.

6 Conclusion and Future Work

Three are the main links one may consider when ranking authors on the Web, namely co-author relationships (AuthorRank aka PageRank), then co-citations which are being progressively provided by digital libraries like ACM or IEEE, as well as recommendations by users for a given article (e.g., through +1-s in Google+).

Alone FOAF social networks are also characterized of links (aka knows) connecting people that could potentially represent authors. Social computing is becoming an indispensible part of the Web including when searching and ranking on the Web.

FOAF has not been designed to support ranking of people. We extended the FOAF ontology with PageRank and AuthorRank metrics: the `hasPR` and `hasAR` properties for each Person (author) individual in the co-authorship network. What is an extreme contribution, we consider, is our innovative approach of designing and implementing the PageRank and AuthorRank algorithms using rules in Semantic Web. Although still work remains to advance this approach to further complex ranking scenarios, our preliminary evaluation results show that there is a promising future to apply that same approach over a wealth of social data on the WWW and their growing in number "social machines".

Our framework may easily be adopted to support ranking social networks in general, given the SWRL rules are then applied to rank people using AuthorRank based on their weighted co-authorship or any other weighted relation, or the reputation gained following, e.g., their `knows` relationships, and the PageRank algorithm instead. The people ranking is growing in popularity since search engines are considering the author's reputation of a Web page when generating search results. Think of Google+ and the FOAF mechanism "behind the scenes" to support author ranking. This is where our future work will concentrate: rank search result of Web pages by considering the reputation of authors of those pages, all supported solely by Semantic Web technologies.

We have developed a framework to support ranking of authors on the Web based on co-author links. As explained, our framework is a semantic AuthorRank algorithm which mimics PageRank. It builds on top of several know ranking metrics and side parameters like are the number of authors, co-authorship exclusivity, PageRank, and co-author weight.

We aim to continue our work on author ranking by considering two other relevant types of inter-author links:

- Co-citation networks (aka h-index, Erdös number, or similar).
- Recommendations by users through social networks (e.g., +1's of Google+, or like's of Facebook).

Acknowledgments We want to thank the personnel of the library at South East European University (SEEU), and the team colleagues of the Linked University Data project funded by the SEEU for providing the set of bibliographic data we experimented with. Also a thank you note is addressed to the IT staff at University of Prishtina for setting up a remote server to run the experiments.

Glossary

ACM (Association for Computing Machinery) A premier membership organization for computing professionals. Unlike the IEEE, the ACM is solely dedicated to computing.

AKT A broad ontology, referred to as the AKT Reference Ontology, for describing an academic computer science community. It consists of several sub-ontologies: Support, Portal, etc.

AuthorRank The PageRank algorithm adopted to rank authors.

Co-authorship networks An important class of social networks modeling co-author relations to, say, analyze the trends of research collaborations, or determine the status of individual researchers.

CO-AUTHORONTO An ontology which extends FOAF to support modeling co-authorship networks on the Web.

DBLP An on-line repository of bibliographic data on major computer science publications well known to its community.

Erdős number An early example of co-authorship networks, in which a "collaborative distance", i.e., the smallest number of co-authorship links between any individual author and the mathematician Paul Erdős is calculated.

FOAF (Friend of a Friend) An RDF schema for machine-readable modeling of homepage-like profiles and social networks.

IEEE (Institute of Electrical and Electronics Engineers) The world's largest professional association dedicated to advancing technological innovation and excellence for the benefit of humanity.

Ontology In computer science, it is a formal representation of knowledge in a given domain of interest through vocabulary definition of that domain: the concepts and their taxonomy, properties, constraints, and relationships among concepts.

OWL (Web Ontology Language) The standard language to formally define and instantiate ontologies on the Web so that they can be used and understood in that certain domain.

PageRank An algorithm based on link analysis for ranking. It was initially used for ranking Web pages on the Web by Google, but is also nowadays used to rank authors (AuthorRank), detect spams, and similar. There is a metric named PageRank in social network analysis (SNA) as well, which measures the reputation of an individual in a network following the same rationale as when ranking Web pages.

RDF (Resource Description Framework) The basic standard for the Semantic Web. It defines the building blocks of the Semantic Web, such as classes and properties, and how they interact to create meaning.

Semantic Web As opposed to the traditional Web, it represents the Web of data *with* meaning in such a way that a computer program can understand (make use of) that meaning of the data.

SNA (Social Network Analysis) An interdisciplinary area of research in social sciences and the information technology. It provides theories and techniques that prove the effects of an individual or a group of individuals belonging to a given network into some outcomes related to that individual or group.

SPARQL (Semantic Protocol and RDF Query Language) The standard language for querying RDF data.

SQWRL (Semantic Query-Enhanced Web Rule Language) A SWRL-based language for querying OWL ontologies.

SWRL (Semantic Web Rule Language) A widely used language to express rules in the Semantic Web.

URI (Uniform Resources Identifiers) Initially used as a standard format to identify pages on the Web (a string usually starting with http://). Today it serves to identify simply anything of interest, be it physical or conceptual, accessible by augmenting it to the Internet. Semantic Web also uses "http" URIs as identifiers for its resources.

References

1. Hendler J, Berners-Lee T (2010) From the Semantic Web to social machines: a research challenge for AI. Artif Intell 174(2):156–161
2. Goldbeck J (2007) The dynamics of Web-based social networks: membership, relationships, and change. First Monday 12(11):1–17
3. Goldbeck J (2005) Web-based social networks: a survey and future directions. Technical report, University of Maryland
4. Brickley D, Miller L (2010). FOAF vocabulary specification 0.97. Retrieved 28 June 2010, from http://xmlns.com/foaf/spec/20100101.html
5. Ahmedi L, Abazi-Bexheti L, Kadriu A (2011) A uniform semantic web framework for co-authorship networks. In: IEEE ninth international conference on dependable, autonomic and secure computing, Sydney, pp 958–965
6. Page L, Brin S, Motwani R, Winograd T (1999) The PageRank citation ranking: bringing order to the Web. Stanford InfoLab
7. Liu X, Bollen J, Nelson ML, Van de Sompel H (2005) Co-authorship networks in the digital library research community. Inf Process Manage 41(6):1462–1480
8. Horrocks I, Patel-Schneider PF, Boley H, Tabet S, Grosof B, Dean M. (2004). SWRL: a semantic web rule language combining OWL and RuleML. Retrieved 27 June 2010, from http://www.w3.org/Submission/2004/SUBM-SWRL-20040521/
9. O'Connor M (2010). SWRLLanguageFAQ. Retrieved from http://protege.cim3.net/cgi-bin/wiki.pl?SWRLLanguageFAQ
10. Ding Y, Scharffe F, Harth A, Hogan A (2006) AuthorRank: ranking improvement for the web. In: Proceedings of the international semantic web and web services conference (SWWS'06)

11. Harth A, Decker S (2005) Optimized index structures for querying RDF from the web. In: Proceedings of the third Latin American web congress. IEEE Computer Society, p 71
12. Erétéo G, Buffa M, Gandon F, Corby O. (2009) Analysis of a real online social network using semantic web frameworks. In: Eighth international semantic web conference (ISWC'09). Wiley
13. De Castro R, Grossman JW (1999) Famous trails to Paul Erdös. Math Intell 21:51–63
14. Nascimento MA, Sander J, Pound J (2003) Analysis of SIGMOD's co-authorship graph. SIGMOD Rec 32(3):8–10
15. The AKT Reference Ontology (2003) Retrieved 23 June 2011, from Advanced Knowledge Technologies. http://www.aktors.org/publications/ontology/
16. Noy N, Rector A (2006). Defining N-ary relations on the semantic web. Retrieved 27 June 2010, from http://www.w3.org/TR/2006/NOTE-swbp-n-aryRelations-20060412/
17. Knublauch H, Fergerson RW, Noy NF, Musen MA (2004). The Protégé OWL Plugin: an open development environment for semantic web applications. In: Third international semantic web conference (ISWC'04.3298). Springer LNCS, pp 229–243
18. McGuinness DL, van Harmelen F (2004). OWL web ontology language overview. Retrieved 27 June 2010, from http://www.w3.org/TR/2004/REC-owl-features-20040210/
19. Carley KM (2001) ORA project. Pittsburg

Detecting Neutral Nodes in a Network of Heterogeneous Agent Based System

Fatemeh Hendijani Fard and Behrouz H. Far

Abstract Software agents with autonomous interaction, negotiation and learning capabilities can be considered as a social network. Central problems in these social networks are: (1) investigating the network to find hot spots (nodes that actively participate in the expansion of the net both physically and functionally); (2) detecting violating nodes, i.e. nodes that have violated certain network policies. Key difference between human and agent based social nets are heterogeneity of agents and having various instances for an agent type. In this paper we propose a technique that can identify neutral nodes, i.e. nodes that can never be considered as a violating or interesting node because it shows similar behavior in all of its interactions. Given the communication protocol and by extracting communication rules, the neutrality for the agent is found in two levels, namely, inter-type and intra-type interactions. The neutrality is found in two steps by giving the network topology and rules of negotiation among agents, and the communication protocol and the extracted communication rules. These two steps guarantee the similarity in frequencies and behaviors of the agents. A direct advantage of this method is reducing the computational complexity for applying analysis techniques and examining all agents to detect a certain property. The techniques are verified through a case study in online auctions. Two additional application areas which widely use agent-based systems are discussed for utilization of the proposed method: e-commerce and e-learning.

F.H. Fard (✉) • B.H. Far
Department of Electrical and Computer Engineering, University of Calgary, Calgary, AB, Canada
e-mail: fhendija@ucalgary.ca; far@ucalgary.ca

Ş. Gündüz-Öğüdücü and A.Ş. Etaner-Uyar (eds.), *Social Networks: Analysis and Case Studies*, Lecture Notes in Social Networks, DOI 10.1007/978-3-7091-1797-2_3,

1 Introduction and Related Works

Multi Agent Systems (MAS) and Social Networks (SN) serve each other in different ways. MAS can be used for agent-based modeling and simulation of social networks [1, 2]. Agent based modeling and simulation models systems with many constituent elements interacting with each other using defined interaction protocols [3]. This approach is required due to the increasing growth of size and complexity of systems [3, 4]. Agent based modeling is particularly useful when modeling systems without a centralized controller. In this case, one can consider the entities participating in a system as agents. The agents can have various types (e.g. dealers in e-commerce systems) and there may be several instances of the same type. There are protocols that govern the behavior by which agents interact with each other. The agent model can help investigate characteristics of the system, observe the changes, and detect emergent behaviors from the interactions. By emergent behavior we mean the unexpected behaviors of the entities which are also mentioned in the studying of emergent behavior in [5, 6].

There has been lots of research using agent based modeling in different application areas. A categorization of different applications of agent based modeling, such as economics and financial markets, crowds, transportation, society and social sciences, biology, etc., is given in [7]. Modeling systems to find the emergent behaviors in the networks regarding timing mechanisms [8], modeling social networks and web service selections in recommender systems [9], modeling migrations in social networks [10], modeling the spread of diseases [11, 12] and mining different measures on groups and teams in social networks [13] are among applications of agent based modeling and analysis.

The agent modeling in a social network context helps extract properties of the network or simulate the agents' behavior in a certain situation. For example, in e-commerce applications, this modeling tries to find the emergent behavior of involved agents and detect possible frauds. In this analysis the agents play the role of members of the social network, their properties such as physical locations, etc. In summary, agent based modeling in this context consists of a set of software agents, their relationship, interaction protocols, network topology, and a framework to simulate the agents' behaviors [7].

On the other hand, agents in a MAS can be viewed as a social network to serve different organizations or systems due to the main characteristics of agents: autonomy, proactivity, and ability to communicate [14]. In this case, the agents can collaboratively achieve an individualized or a group goal [15, 16]. The social network rules and algorithms can be applied on the participating agents to analyze MAS from the social network perspective [17, 18].

In this paper, the network of agents is investigated where there are agents that may violate the rules or develop an emergent behavior. For example, in e-commerce or e-learning systems, the agents can act on behalf of real individual or organizations. Therefore, the agents' network and communication should be studied to detect fraud, identify trusted agents, find creative agents, etc. There has been researches that detect the emergent behavior (or better known as anomalies in the social networks) using agent based simulation or proposing different criteria and algorithms [5, 19–22]. However, approaches developed specifically for networks of software agents considering the agents' concepts are rarely found. A challenge in the study of social networks of agents is the large number of agents, heterogeneity of agents, having various instances of the same type, and complexity of the agents' communications [23].

In this paper we focus on a technique to reduce the complexity of analysis in a network of heterogeneous software agents. The proposed method suggests detecting neutral agents, i.e. the agents that behave with similar frequencies and show the same behavior in their interactions with other agents. We argue that through identifying and eliminating these agents the complexity of the network and the algorithms to analyze them will be decreased.

For detecting neutral agents both inter and intra relationships of the agents are examined. The inter-type relationships show the interactions of different agent types, and the intra-type considers the communications between instances of the same type. The agents' communications can be classified into inter- or intra- type communication and then be examined by each of the modules in the designed system. For example, in e-commerce online auctions, agents can be of Seller or Buyer types, while there are several sellers or buyers interacting in the system i.e. instances of Seller or Buyer type. The interaction between Seller and Buyer types are examined in one level and interactions of sellers or buyers in the other level.

We have developed a technique used for inter-type interactions and performed a case study on e-commerce, in which there are certain types of individuals collaborating with each other in specific roles. These roles can be considered as agent types. Then through an agent modeling we have investigated the agents that are neutral in relation with other types.

The other advantage of the proposed technique besides reducing analysis complexity, is considering the applied rules defined on the communication of different agents. This is also the originality of our technique. The detection of neutral agents cannot be done with the existing data mining techniques; since they mine data without considering the interaction protocols and the rules defined for each agent. However, these protocols are the basics of our technique to detect the neutral agents.

The structure of this paper is as follows: in Sect. 2 the proposed method is explained in detail. Section 3 describes the technique verified through a case study in a multi agent online auction system. Application areas are defined in Sect. 4 followed by discussions and conclusion parts in Sect. 5.

2 Proposed Method

We investigate the agents' communication in two different levels: inter and intra relationship of agents. In each level, the neutral nodes are detected based on the analysis of agents' communications and communication protocols.

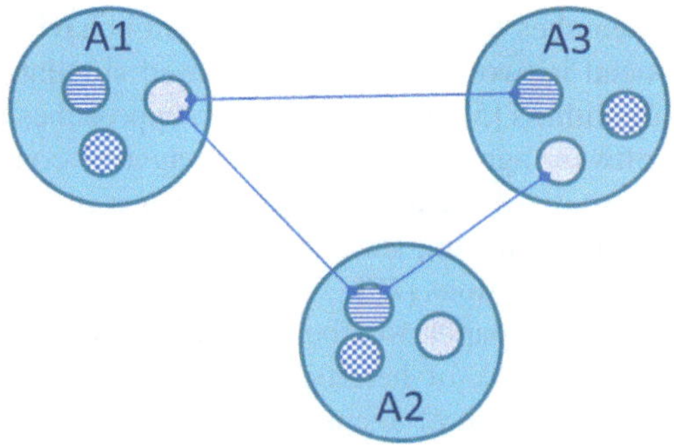

Fig. 1 Agents' communication with specific roles

The method is based on the interaction of agents through their roles and the assigned tasks to each role. In the methodologies used for developing agent based systems, each agent can have multiple roles and each role performs specific assigned tasks. For example in an irrigation system, a water control agent can have two different roles: analyzer role and balancing role. The analyzer communicates with water sensors, other agents, and databases to analyze how much water for each part of the greenhouse is needed. The balancing role on the other hand, performs the tasks related to balancing mist and moisturizing. For each of the assigned tasks, the agent interacts with different roles of other agents. Figure 1 is an illustrative example of three agents A_1, A_2, and A_3 in a network. Each agent has three roles. The agents communicate with each other based on the tasks assigned to each role and protocols defined for communication.

The proposed method's architecture is shown in Fig. 2. The inter-type relationship modules are shown in the left part of the figure and the modules for detecting neutral nodes in intra-type relationship are shown in the right. The inter-type component consists of two modules for inspecting interactions between various types of agents: detect similar frequencies and detect similar behaviors. Depending on the application and/or mining the agents' network, one or both of them can be used for detecting neutral agents in this level; i.e. the agent types that show same or neutral behavior in their communication with other agent types. The intra-type relationship component searches interactions among instances of one type of agent. This component consists of two modules for detecting neutral nodes: matching roles and detect similarities.

Both components use repository of protocols and data logs of the agents' network. Protocols show the rules defined for the interaction of agents to achieve individual or system (organization) goals for the agents. These interactions are defined for agents in agent based modeling and simulation, or can be in the form of UML's Sequence Diagrams, scenarios specification languages, or other formats based on the methodology used for MAS design. The detailed explanations of the components are defined next.

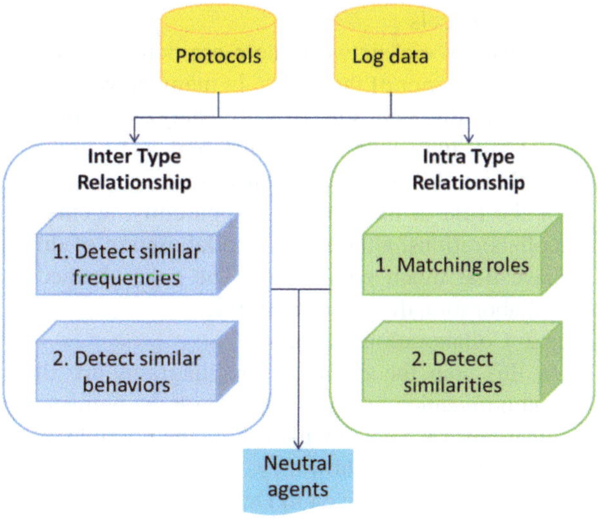

Fig. 2 System components for detecting neutral agents

2.1 *Inter-Type Component*

Two modules comprise the inter-type component. The first module detects the similar frequencies in the behaviors of the agent in its different communications. The second one considers the roles and tasks an agent performs in relation with other agents. The system has been designed in a modular way, and one or both of these modules can be used for the detection of neutral agents based on the user needs and problem types.

2.1.1 Detect Similar Frequencies

Software agents are autonomous software entities that interact with each other in a computing environment. Each agent has its own set of operations to perform. It also has a model of interactions with the other agents [24, 25]. The interaction rules are defined in the artifacts of an agent based software engineering methodology (AOSE) that is used to develop the agent-based system. The interaction rules are referred to as protocols in this paper. The "detect similar frequencies" module supports the case when each agent plays a single role to achieve its goals. This method has the following steps considering both cases:

Step 0: Agent types

In this step different agent types that correspond to different nodes of the social network are identified. Each agent type will be associated with a set of operations that each node can perform in the network. For example in the case study below, *send a message* and *share* are operations available to *User* agent types and *delete message* is action available to *Moderator* agent type.

Step 1: Protocols

For each social network there are defined rules of interactions that govern the social behavior of the nodes. In many networks the rules are given in declarative form. In this step, the rules are analyzed and converted to the procedural forms that are applicable to the different agent types. Each rule is modeled using UML's collaboration diagram among agent types that represents implementation of the rule in the network. Apparently, agents of one type can have complex collaborations, although the collaboration of different types of agents may not have that much complexity. The rules defined here help us to illustrate the collaboration diagrams in Step 2.

Step 2: Collaboration diagrams

In this step the collaboration diagrams are illustrated based on the agents' communication protocols. Based on the extracted information, the collaboration diagrams are created which show the possible interaction path between different types of agents and how they can communicate with each other. Definition related to this diagram is given in Definition 1.

The advantage of using collaboration diagrams is that the complex social network which shows the relationship between all the agents is reduced to a set of graphs. This shows the relationship between different types of agents and helps reduce the complexity of analyzing the social networks. Also the difference of the collaboration diagrams with protocols is that these diagrams just show the number of messages and unlike protocols they don't contain the message contents.

Step 3: Collaboration matrices

In this step, the relationships between different agent types are transformed to collaboration matrices explained in Definition 2.

Step 4: Clustering the behaviors

After constructing the matrices the clustering algorithm with Manhattan Distance [23] is applied to find the behavior of an agent type in different collaboration diagrams.

Step 5: Detect neutral agents

In this step, the results from previous step are analyzed and the agents which are neutral are extracted. The neutral agents show the same behavior in all of the possible collaborations. It means that they have a single behavioral pattern when interacting with the other agent types. Therefore the relations between this neutral agent and other agent types may not cause any changes to the network. The algorithm and definitions are explained below.

Definitions and Algorithm

In this section, the related definitions are explained. Definitions for collaboration diagrams, collaboration matrices and vectors related to each agent are as follow:

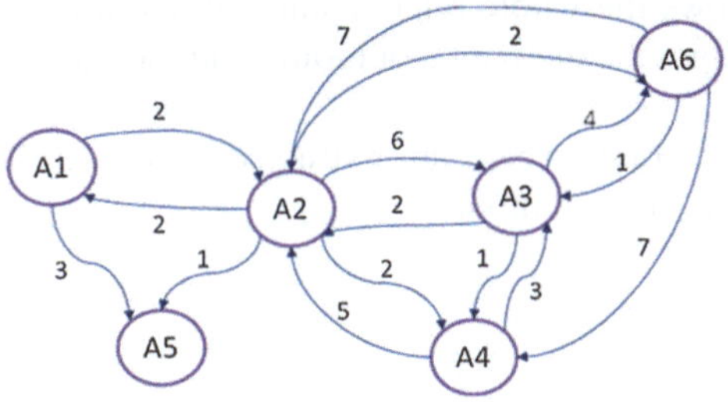

$$M1 = \begin{array}{c|cccccc} & A1 & A2 & A3 & A4 & A5 & A6 \\ \hline A1 & 0 & 2 & 0 & 0 & 3 & 0 \\ A2 & 2 & 0 & 6 & 2 & 1 & 2 \\ A3 & 0 & 2 & 0 & 1 & 0 & 4 \\ A4 & 0 & 5 & 3 & 0 & 0 & 0 \\ A5 & 0 & 0 & 0 & 0 & 0 & 0 \\ A6 & 0 & 7 & 1 & 7 & 0 & 0 \end{array}$$

Fig. 3 An example of a collaboration diagram with six agent types and its collaboration matrix

Definition 1 A collaboration diagram is a directed graph $G=(V, E, W)$ in which the vertices, denoted by $V(G)$, are agent types. In a collaboration diagram $|V(G)|$ equals to the total number of agent types. The edges, denoted by $E(G)$, are relationship between agent types. In graph G there is an edge $e_{ij}=(v_i, v_j)$ from v_i to v_j if there is a message sent from v_i to v_j. $W(G)$ shows the weights of the edges $E(G)$. For each edge e_{ij}, $w(e_{ij})$ is determined by the number of messages sent from node i to node j. The messages correspond to allowable messages defined in the interaction protocols.

Definition 2 A collaboration matrix M for a collaboration diagram G is a square matrix $M=[m_{ij}]_{n\times n}$ where n is the number of vertices in graph G, i.e. $n=|\mathrm{V(G)}|$. In matrix M, an entry m_{ij} is non-zero if there is a corresponding edge $e_{ij}=(v_i, v_j)$ in the related graph and the value of m_{ij} equals to the weight of that edge, $m_{ij}=w(e_{ij})$. Otherwise $m_{ij}=0$.

Definition 3 A vector vec_{ak} related to an agent type A_k in a collaboration diagram G_a shows the whole incoming and outgoing edges of that agent in the related collaboration diagram. This vector can be derived from the collaboration matrix M_a. Therefore, a vector vec_{ak} for an agent A_k related to a collaboration matrix M_a is defined as $vec_{ak}=[row_{ak}, (col_{ak})^T]$ where row_{ak} is the entries of kth row of related matrix and $(col_{ak})^T$ is the transpose of the kth column of that matrix. The kth row in matrix M shows the outgoing messages of agent A_k of and the kth column shows the incoming messages from other agents to agent A_k in the related collaboration diagram.

An example of a collaboration diagram composed of six agent types is shown in Fig. 3. The collaboration matrix for this diagram is shown in matrix M1. In this matrix, for example, agent A_2 has sent messages to agents A_1, A_3, A_4, A_5, and A_6. This is clarified by considering the entries in second row. Also, the incoming edges for agent A_2 or the received messages can be defined, too. This is obtained by considering the second column. Therefore, for this agent the messages are received from agents A_1, A_3, A_4, and A_6.

For the diagram shown in Fig. 3 the vector which shows the relationships of agent A_2 in matrix M1 is defined as $\mathrm{vec}_{12}=[\mathrm{row}_{12}+(\mathrm{col}12\ )^{\mathrm{T}}]$. The vector $\mathrm{row}_{12}=\{2, 0, 6, 2, 1, 2\}$ and $(\mathrm{col}_{12})^{\mathrm{T}}=\{2, 0, 2, 5, 0, 7\}$. Therefore we have $\mathrm{vec}_{12}=\{2, 0, 6,$

2, 1, 2, 2, 0, 2, 5, 0, 7}. This vector shows the whole relationships of agent A_2 in collaboration diagram G_1 with other agents (the incoming, outgoing and the number of messages in their communication).

The agent types and collaboration diagrams which are extracted based on the interaction protocols are the detection algorithm's inputs.

Algorithm 1 Detect Similar Frequencies

Input: number of agent types (n), d collaboration diagrams
in the form of G= (V, E, W)
Output: list of neutral agents with similar frequencies in inter-type level
1. Transform collaboration diagrams into d collaboration matrices $M_a = [m_{aij}]_{n \times n}$ where $1 \le a \le d$
2. For each agent type do the following
3. Make all the related vectors of the *d* matrices in the form of $vec_{ak} = [row_{ak}, (col_{ak})^T]$ where $1 \le a \le d$ and $1 \le k \le n$
4. End for
5. For each of the *n* agent types do the following
6. If vector vec_{ak} equals to zero
7. Eliminate that vector
8. End if
9. Cluster all nonzero vectors related to that agent with Manhattan Distance
10. End for
11. For each of the *n* agent types
12. If the vectors remain in the same cluster
13. Print that agent type as a neutral agent
14. Else
15. Print the number of clusters for each agent type and the cluster numbers for each collaboration diagram
16. End for

In *the first step of the algorithm* the collaboration matrices (Definition 2) are extracted from the collaboration diagrams. All the relationships shown on a collaboration diagram are transferred to the collaboration matrices. These matrices, as mentioned before, show all the incoming and outgoing messages from one agent to the other agents and vice-versa. Next, the agent types are clustered into different groups. Each agent type is clustered based on its relationship in different networks with other agents. For this purpose, all the vectors related to that agent are extracted from the matrices. Clustering is applied on the non-zero vectors related to each type of agent. Zero vectors should be omitted because they show that this agent is not participated in a collaboration diagram. In other words, the existing relationships are important and therefore zero vectors are omitted (*steps 2–10 of the algorithm*). By this clustering one can understand the changes that exist in the relations of one agent to other agents in different networks. If there is no change in this relationship, then the agent is neutral in communication with other types (*steps*

11–16 of the algorithm). Therefore these neutral agents are omitted in the analysis of inter relational behavior of different roles in social networks and can be excluded from the analysis phase.

2.1.2 Detect Similar Behaviors

The second module for the inter type relationship component (see Fig. 2) is detecting similar behaviors in between the communications of different agent types. As mentioned before, each agent has specific tasks to perform which is assigned to different roles of the agent (see Fig. 1). Therefore, for detecting similar behaviors the agents' network is investigated based on their assigned roles and tasks. Similar behaviors here are detected by the agent's communications and tasks through its roles. This module investigates the real behavior of each agent type considering its roles, performing tasks, and the set of interacting agent types. The steps involved in this module are:

Step 1:

The instances of each agent type are clustered based on their roles and their agent types' communication list. For example, for agent type *Z* the assigned roles are a set of *roles* = $\{r_{zi} | z = \text{agent type } Z, i \in [1, \# \text{ of roles of agent type } Z]\}$.
All instances of agent type *Z* are clustered to show each role is communicating with a set of agent types $T = \{T_t | t \in [1, \# \text{ of Agents types}]\}$ to perform a set of *tasks* = $\{\text{tasks}_{zi} | z = \text{agent type } Z, i \in [1, \# \text{ tasks}]\}$. This data is extracted from the log data which shows the real data communicated in the network.

Step 2:

Extract the communication rule based on the clustering from previous step in the form of a four-tuple $(A_z, r_{zi}, \text{tasks}_{zi}, T)$ where:

- A_z: shows the agent of type *Z*
- r_{zi}: shows i^{th} role of A_z
- tasks_{zi}: shows a set of tasks A_z can perform through its ith role
- T: shows a set of agent types $\{A_m, \ldots, A_p\}$ that A_z communicates with its ith role to perform tasks_{zi}

Step 3:

For one agent type, the agent is reported as a neutral agent if:

- The extracted rules of step 2 matches the protocols; and
- The set of tasks each agent performs by a specific role remains the same in communication with a specific set of agent types T

If both conditions are satisfied, the agent has shown similar behavior based on its roles in its communication with other types. These agents may not be a point of emergence and can be reported as neutral agents. The algorithm for performing steps 1–3 is described in next.

Algorithm 2 Detect Similar Behaviors

Input: agent types, agents' protocols for each agent type in the form of (A_z, r_{zi}, $tasks_{zi}$, T), log data
Output: list of neutral agents with similar behaviors in inter-type level

1. For each type of agent do the following
2.   Find the set of communication protocols for this type in the form of (A_z, r_{zi}, $tasks_{zi}$, T)
3.   For each role of the agent do the following
4.   Extract the communication rules from its log data in the form of (A_z, r_{zi}, $tasks_{zi}$, T)
5.   If the communication rule match the protocol
6.   If there is one communication rule for each set of role and tasks
7.   Report the agent as neutral node
8.   End if
9.   End if
10.   End for
11. End for

2.2 *Intra-Type Relationship*

In this section, the right part of Fig. 2 is explained. The agent instances of each type are studied to detect the neutral nodes in their intra network.

Each agent for performing specific tasks is in interaction with other agents through its roles. By identifying the communicating roles of two agents for a set of tasks, the agent behavior is identified. In Fig. 1 the specific roles of the three agents of one type are shown with different fill patterns. Each agent interacts with one or more of other agents' roles through its roles. These interacting roles and the assigned tasks are defined in the protocols in the form of ($role_i$, $tasks_j$, $role_k$) where:

- $role_i$: shows the ith role of one agent
- $tasks_j$: shows the set of tasks that the agent performs through its ith role when interacting with $role_k$ of another agent of its own type
- $role_k$: shows the kth role of another agent with the same type

Also, the communicating roles show the agent's behavior in the network. By finding the behavior of each agent and match it with the protocols, one can find whether the agent is a neutral node or not. Likewise the previous module, if the agent shows the same behavior it is considered as a neutral node.

This module is similar to the "detect similar behaviors" module. So a brief algorithm showing the steps and not the implementation details is explained:

Algorithm 3 Detect Neutral Agents in Intra Type Level

Input: agents' communication protocols in the form of ($role_i$, $tasks_j$, $role_k$), log data
Output: neutral agents in intra-relationship level
12. For all agents of one type do the following
13. For each agent do the following
14. Consider the set of its protocols in the form of ($role_i$, $tasks_j$, $role_k$)
15. Find all of the communication roles for each agent from its log data in the form of ($role_i$, $tasks_j$, $role_k$)
16. If these communication roles match the protocols
17. If the matched communication roles are the same in all its communications
18. Report the agent as neutral node
19. End if
20. End if
21. End for
22. End for

3 Case Study

As a case study, an e-business social network implementing an e-commerce system is considered. This is an open auction agent based system in which seller and buyer agents can participate and provide goods and/or bid on goods. In this system, learning or changing in the behavior can happen [24]. It is assumed that agents are rational, with defined objectives (e.g. maximizing their individual utility) and follow rules of behavior as decided by the auction hosting authority [7]. The agents, such as buyers and sellers, can show new behaviors based on their previous experiences and competition factors which exist for each of them [25]. An auction system consists of multiple agents with different types that communicate with each other autonomously, while most of the agents are of the same type.

Modeling and analysis of such a system is important regarding the fraud detection or other trust and security issues. In this case study it is shown how the proposed method can reduce the complexity related to analyzing such a network and analyze the relationship between different types of agents before analyzing the whole network.

For explanation simplicity six types of agents are considered here:

- Controllers or market place agents (C or A_1): These agents are responsible for keeping the auction types, the protocol for each auction, different protocols for each auction and other related rules.
- Auctioneers (A or A_2): Agents responsible for taking the auction actions. Declaring the protocol of auctions, declaring the sellers, buyers, and items in each auction, taking the time, accepting bids from the buyers, and deciding about the winner of an auction.

- Registrars (R or A_3): Agents responsible for the communications between the sellers and buyer with the auctioneer. These agents introduce the sellers and buyers to the auctioneers, register new users, and take the log in functions. Also the search for each item or auction type is passed to these agents.
- Sellers (S or A_4): Agents which have some items and decide to sell their items in each type of auction. Sellers should register first to the registrar auction for selling the items. These agents declare the items and their first price to the registrars.
- Buyers (B or A_5): These agents look for items to buy and bid for each item they want. Buyers should register first and then log in each time they want to participate in an auction.
- Trust managers (T or A_6): These agents assess and represent trust criteria for other agents taking part in online auctions. The trust managers help the other agents in decision making.

There are other agent types like the Credit Associate (CA) which are responsible for the payment options for sellers and buyers. The other types are not mentioned for simplicity.

The relationship between these agent types is defined with scenarios that indicate the behavioral protocols for participant agents. For example, for registering a new buyer into an auction system, a message is sent from the buyer to the registrar agent. The request contains the buyer's information. A few messages are communicated between the two agents e.g. security question, and user name and password, and after a few messages the buyer is registered. The same procedure is followed when a new seller wants to register with the system. In either case, the information of auctions is passed from the auctioneer to registrar and shown on the website. This process is named scenario 1 for registering a new user into the auction system.

A few of these scenarios are named in the below list:

Scenario 1: As explained above, registering a new user (buyer or seller agent) into the auction system is done in this scenario.

Scenario 2: In this scenario the registered users (either sellers or buyers) try to sign in based on the information declared on the website about the auctions.

Scenario 3: The registrar agent registers seller and buyer agents into a certain auction. The registrar agent then introduces them to the auctioneer agent.

Scenario 4: The registered users in a certain type of auction take part in this scenario. The actions related to performing an auction e.g. bidding, declaring the winner, and introducing the sold items information are done in this scenario. The trust manager agent also participates and has an important role here. The protocols and criteria found and described to calculate the amount of trust to other agents are executed by the trust manager agents. The messages between different seller and buyer agents to the trust manager agents are defined in this scenario.

Scenario 5: In this scenario the rules on how the buyer agents can search for items are defined.

Scenario 6: In an auction system seller agents can search for certain types of auctions and get the related information and protocols of each auction. The related communication rules are defined in this scenario.

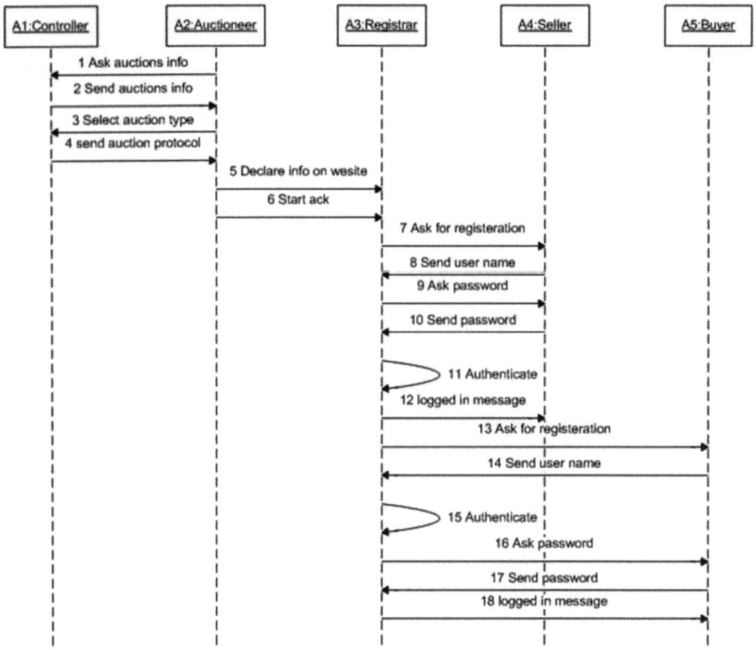

Fig. 4 Users sign in based on the auctions information on the website

Scenario 7: In this scenario, the seller agents can declare the items they want to sell, prices of each of them, and the type of auction to the registrar. The registrar agent puts this information on the website.

This set of scenarios is open and expandable. In real world, there are more scenarios that can be defined for the communications between different types of agents. An advantage of the method proposed in this paper is that it allows seamless inclusion and integration of new scenarios and new agent types.

For better understanding the communications between agent types and the rules on the order of messages, sequence diagrams, as shown in Fig. 4, can be used. In this figure each vertical line represents an agent instance. The communications are presented in the form of messages and shown with arrows from one agent to another. The message labels are written above the arrows. The order of messages communicated between the agents is shown with order of arrows in the vertical line. It means that the first messages are upper than the other ones or the timeline starts from top of the vertical lines and ends at the bottom. Figure 4 shows the communications of agents related to Scenario 2.

The third scenario is shown in Fig. 5 where the registrar agent registers seller and buyer agents into a certain auction.

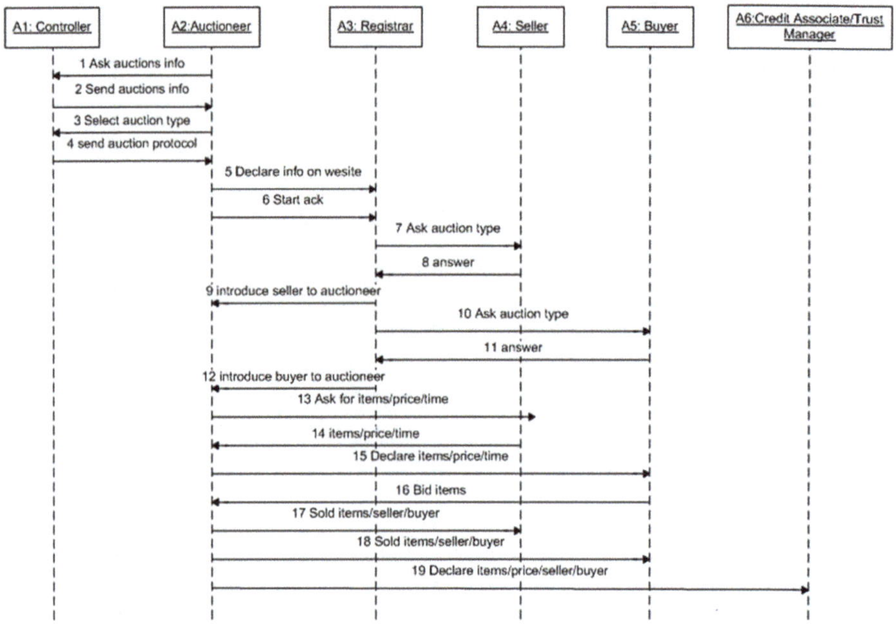

Fig. 5 Registering existing users for a certain auction

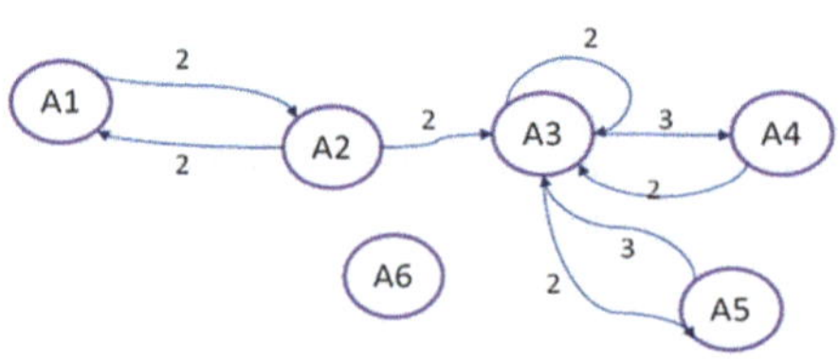

$$M2 = \begin{bmatrix} & A1 & A2 & A3 & A4 & A5 & A6 \\ A1 & 0 & 2 & 0 & 0 & 0 & 0 \\ A2 & 2 & 0 & 2 & 0 & 0 & 0 \\ A3 & 0 & 0 & 2 & 3 & 3 & 0 \\ A4 & 0 & 0 & 2 & 0 & 0 & 0 \\ A5 & 0 & 0 & 2 & 0 & 0 & 0 \\ A6 & 0 & 0 & 0 & 0 & 0 & 0 \end{bmatrix}$$

Fig. 6 Collaboration diagram related to Scenario 2

3.1 Detecting Similar Frequencies Module for Inter-Type Relationship

For applying the modules explained in the previous section, we start with the inter-type relationship module for detecting similar frequencies. Therefore, it is needed to extract the collaboration diagrams and matrices. The collaboration diagram of Scenario 2 is shown in Fig. 6. This diagram is produced using the Definition 1. The collaboration matrix made from the collaboration diagram of Scenario 2 is shown in matrix M2.

Some of the collaboration diagrams for other scenarios are shown in Figs. 7 and 8 that together represent some possible interaction networks for this system.

By applying *Algorithm 1*, it is concluded that the controller and trust manager agents (Agents 1 and 6) remain in the same cluster. It means the communication

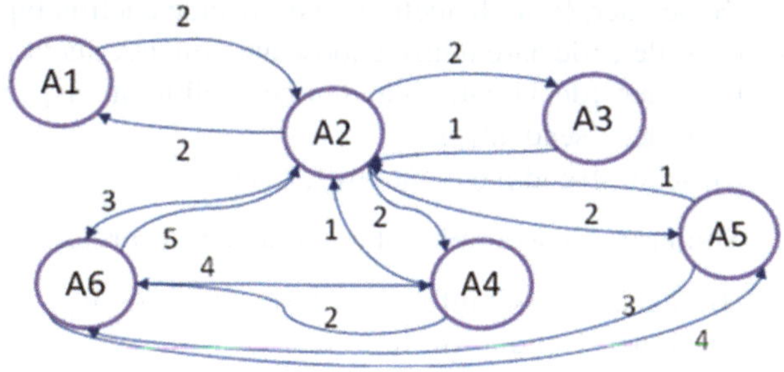

Fig. 7 Collaboration diagram of Scenario 4

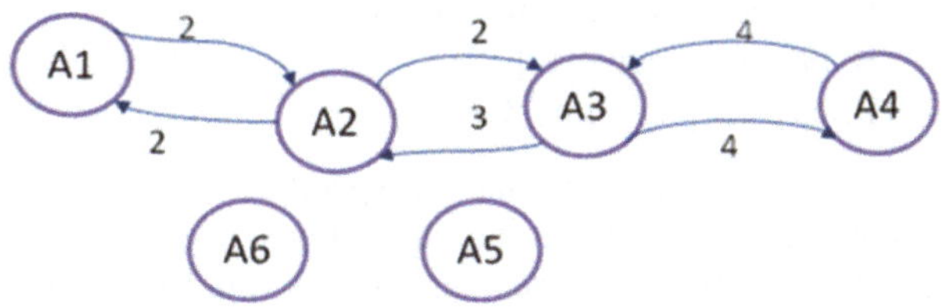

Fig. 8 Collaboration diagram of Scenario 7

of these two agents with other agents in the given networks will show the same behavior. For controller agent (A_1) the vectors in the seven scenarios are extracted from the collaboration matrices of the diagrams by Definition 3. From this definition, the vectors related to this agent in all networks have the format: $vec_{a2} = \{0, 2, 0, 0, 0, 0, 0, 2, 0, 0, 0, 0\}$ where $1 \leq a \leq 7$. The results of clustering the related vectors show that the seven vectors are in the same cluster.

For trust manager agent (A_6) again we have $vec_{a6} = \{0, 5, 0, 4, 4, 0, 0, 3, 0, 2, 3, 0\}$ for $a = 3, 4$ and $vec_{a6} = \{0, 0, 0, 0, 0, 0, 0, 0, 0, 0, 0, 0\}$ for $a = 1, 2, 5, 6, 7$. Following the Algorithm 1, the non-zero vectors should be considered. Therefore again the results show that this agent remains in the same cluster and it shows the same behavior in all the networks it has communication with others.

3.2 *Detecting Similar Behaviors Module for Inter-Type Relationship*

The second module, detects similar behaviors in the inter-type relationship. For this case, two agent types Seller and Buyer are considered. To show the applicability of the proposed technique, the example here shows agent types that can't be consider as a neutral agent. Agent types S, B, R, A, CA represent agent types Seller, Buyer, Registrar, Auctioneer, and Credit Associate respectively.

The communication rule or communication protocol of each type is defined as a 4 tuple (A_z, r_{zi}, {$tasks_{zi}$}, T). The Seller type can have the following communications:

- (S, registrant, {set info, set security options}, R)
- (S, signer, {send info}, R)

- (S, decider, {search auction types, decide auction type}, R)
- (S, seller, {declare items, choose auction, ask auction protocols}, R)
- (S, seller, {declare items and prices, selling items}, A)
- (S, seller, {send ad}, B)
- (S, seller, {send payment info}, CA)

Similarly, the examples of communication protocols for agent type Buyer are:

- (B, registrant, {set info, set security options}, R)
- (B, signer, {send info}, R)
- (B, searcher, {search auction types, search items}, R)
- (B, buyer, {choose items, bidding}, A)
- (B, buyer, {receive ad}, S)
- (B, buyer, {payment}, CA)

Each agent type can have a specific role with its set of assigned tasks when communicating with another agent type. A simple communication that can happen as a fraud is that the *Seller* changes its role for the *Buyer* and asks its information. This can have a communication like (S, seller, {send payment info}, B). The other problem can be a communication like this: (S, buyer, {choose items, bidding}, A) when *Seller* tries to bid its own item or (B, seller, {send payment info}, CA) when the *Buyer* tries to fake its payment information as a *Seller*.

The proposed Algorithm 2 does not report these agents as neutral ones. Because in the "if" conditions they don't satisfy either matching with their protocols, or showing the same rule in all of their interactions.

3.3 Detecting Similar Behaviors in Intra-Type Relationship

The allowed intra-type relationship of the agents can be in the form of ($role_i$, $tasks_j$, $role_k$) which shows the legal tasks set to be performed between roles. Continuing the example, one of the protocols for agents of type *Seller* is (finder, {ask selling items, seller info}, representer), where the sellers can ask regular information from another seller.

One of the extracted rules for one seller can be (finder, {selling items, seller info}, representer) or (finder, {price negotiation}, representer). While the former matches the protocol, the latter doesn't match any of the protocols and so is not reported as a neutral agent.

For another agent, if just the first rule is extracted, since it matches the protocol it satisfies the first condition of Algorithm 3. However, it should be examined if this rule remains the same in all of the agent's communications. If this condition is satisfied too, the agent is reported as a neutral node.

This case study shows that two types of agents have similar frequencies in collaboration with other agent types based on Algorithm 1. For example, the trust manager agent's communications with other five agent types is similar in the diagrams that it appears in. In other words, it shows a neutral behavior in the networks and in communication with the other agents. Therefore, it is unlikely that this agent changes the model of network in future and show a new behavior. The reduction of two agent types means approximately 33 % decrease of the size of different types in this example for further analysis. Note that this conclusion is dynamically evolvable when new agent types and scenarios are added to the system.

For Algorithms 2 and 3, we verified the correctness of the technique by focusing on the conditions in these algorithms. However, if the conditions are satisfied, the agents have shown similar behavior in their interactions and are reported as neutral agents.

The problems arising when communication rules changes can be further analyzed for security or fraud issues. The proposed technique in this article can be used as a preliminary technique for detecting such issues when analyzing a network of agents for the changes in the roles or tasks of each agent communicating with other types against its permitted protocols.

4 Application Areas

The techniques developed in this paper can be applied in various fields. Two of the main application areas, e-commerce and e-learning [26] application are discussed in this section. In both areas, heterogeneous and open agent based systems can be used.

One possible application of the technique presented here is e-learning. In e-learning two problems have taken a lot of attention in recent years: detecting experts, and finding creative people. For investigating such people, usually the characteristics of networks and the incoming and outgoing relations of one node are investigated. When agents play role in this area, the neutral nodes can be omitted to improve the complexity problem. Omitting such agents especially for detecting creativity can be helpful. Because neutral nodes detected with our technique, have same behavior in all their interactions and don't show creativity behaviors based on the creativity characters mentioned in related works.

In e-commerce applications there is a broad demand on the security issues and detection of fraud. The technique proposed here can be considered as a filtering by detecting neutral agents in a network of software agents. Also, it can detect the possible fraudulent agents by modifying the proposed algorithms to detect the agents that violate communication protocols or show changes in their set of roles, and tasks each type performs. Furthermore, considering two levels for the agents' interactions can focus on the details on interactions to detect the exact point of emerging new behaviors. In intra-type relationship, if the set of roles changes, or this set does not match the protocols, the agent can be a fraudulent agent showing unusual behavior. For example when a buyer interacts with other buyers as a seller or sellers negotiate on items' prices known as phishing.

The other possible application is the detection of agents that change their role in their life cycle. For example, an agent in e-learning can change its role from a Learner to a Teacher after a learning period. Based on this role changing, the authorizations should be changed too. Applying the developed technique can avoid the unauthorized usage when agents play different roles in their life time.

5 Discussion and Conclusion

There are numerous communications in the social networks and a large number of them are the communications between instances of one agent type, while others are the communications between different types of agents. Although intra-type communications are possible in a generic social network, they are governed by rules of interaction in the network. For example, in the auction system, communication between individual buyers or sellers may be prohibited. However, mining the interactions between heterogeneous software agents can be complex due to different concepts, protocols, roles, and tasks each type perform. This analysis becomes more complex when mining all agents' interactions and when the number of participating agents grows.

The method proposed in this paper detects neutral agents in a social network with respect to inter-type and intra-type communications. It suggests eliminating neutral agents from later stages of modeling and analysis since they don't show new behavior in networks. Therefore, they are unlikely to change the topology of network or cause the whole network to change over time. Since these agents that are known as neutral agents show similar frequencies and similar behaviors in their interactions.

Consequently, by finding the neutral agents, we can have more focus on other agents and also reduce the complexity or time and cost of analyzing the networks. This reduction depends on the relationships and number of agent types in each network. The key point here is investigating the inter-type and intra-type communications separately. This classification can help having more focus on each level of interactions and identify the agent's behavior in each level. One agent can be found as a neutral node either in its inter- or intra- relations or both. The other advantage of this classification is reducing complexity of the interaction analysis.

Also it should be mentioned that the results of clustering in Algorithm 1 can be helpful in next steps of analysis and modeling of the social networks. For example, in the case study the result of clustering shows same clusters for some agents such as agents 2, 4 and 5 (Auctioneers, Sellers and Buyers) in some of the participating networks. Results of such clustering can be helpful in analysis of networks since it shows in which possible networks the behavior of one type of agent is similar and they can be treated almost the same in the modeling. This can be pointed out as a future research to work on.

Also it is found that the agents that have more number of clusters have a higher possibility to show an emergent behavior in the network, since they have a new behavior in each possible network.

Other possible future directions include the improvement of algorithms by considering the agents with the messages in their collaborations. This can be helpful in modeling and analysis of social networks by their semantics.

Acknowledgment This research was partially supported by a grant from Natural Sciences and Engineering Research Council of Canada (NSERC).

References

1. Frias-Martinez E, Williamson G, Frias-Martinez V (2011) An agent-based model of epidemic spread using human mobility and social network information. in privacy, security, risk and trust (passat). In: IEEE third international conference on social computing (SOCIALCOM)
2. Taveter K, Du H, Huhns MN (2012) Engineering societal information systems by agent-oriented modeling. J Ambient Intell Smart Environ 4(3):227–252
3. Macal CM, North MJ (2006) Tutorial on agent-based modeling and simulation. Part 2: How to model with agents. In: Proceedings of the winter simulation conference 2006 (WSC 06)
4. Yang H et al (2009) Multi-agent based modeling and simulation of complex system in hospital. In: Sixteenth international conference on industrial engineering and engineering management 2009 (IE&EM'09)
5. Haglich P, Rouff C, Pullum L (2010) Detecting emergence in social networks. In: Proceedings of the IEEE second international conference on social computing 2010, pp 693–696
6. Fard FH, Far BH (2012) Detecting emergent behavior in autonomous distributed systems with many components of the same type. In: IEEE international conference on systems, man, and cybernetics (SMC)
7. Macal CM, North MJ (2008) Agent-based modeling and simulation: ABMS examples. In: Winter simulation conference 2008 (WSC 2008)
8. Seung Man L, Pritchett AR (2008) Predicting interactions between agents in agent-based modeling and simulation of sociotechnical systems. IEEE Trans Syst Man Cybernet Part A 38(6):1210–1220
9. Al-Sharawneh J, Williams MA (2009) ABMS: agent-based modeling and simulation in web service selection. In: International conference on management and service science 2009 (MASS'09)
10. Filho HSB, de Lima Neto FB, Fusco W (2011) Migration and social networks—an explanatory multi-evolutionary agent-based model. In: IEEE symposium on intelligent agent (IA)
11. Yuanzheng G et al (2011) Agent based modeling for H1N1 influenza in artificial campus. In: Second IEEE international conference on emergency management and management sciences (ICEMMS)
12. Zhang F, Zhao Q-x, Li L (2011) Intervention for contagious disease: agent-based modeling and simulation. In: Second IEEE international conference on emergency management and management sciences (ICEMMS)
13. Do Young C, Young Wook S, Kun Chang L (2011) Agent-based modeling approach to a team creativity pattern analysis based on heterogeneity and network structure. In: International conference on management and service science (MASS)
14. Fortino G, Palau CE (2012) An agent-based mobile social network. In: International conference on multimedia computing and systems (ICMCS)

15. Bergenti F, Franchi E, Poggi A (2011) Agent-based social networks for enterprise collaboration. In: 20th IEEE international workshops on enabling technologies: infrastructure for collaborative enterprises (WETICE)
16. King Lun C, Wing Bun L (2001) Multi-agent based virtual enterprise supply chain network for order management. In: Portland international conference on management of engineering and technology 2001 (PICMET'01)
17. Bentahar J, Khosravifar B, Gomrokchi M (2009) Social network-based trust for agent-based services. In: International conference on advanced information networking and applications workshops 2009 (WAINA'09)
18. Enembreck F, Barthès J-PA (2013) A social approach for learning agents. Expert Syst Appl 40:1902–1916
19. You C et al (2011) Leveraging social networks to detect anomalous insider actions in collaborative environments. In: IEEE international conference on intelligence and security informatics (ISI)
20. Yongkun L, Lui JCS (2011) Friends or foes: detecting dishonest recommenders in online social networks. In: Proceedings of 20th international conference on computer communications and networks (ICCCN)
21. Benevenuto F et al (2009) Detecting spammers and content promoters in online video social networks. In: IEEE INFOCOM workshops
22. Enhua T et al (2012) Spammer behavior analysis and detection in user generated content on social networks. In: IEEE 32nd international conference on distributed computing systems (ICDCS)
23. Fard FH, Far BH (2012) Clustering social networks to remove neutral nodes. In: IEEE/ACM international conference on advances in social networks analysis and mining (ASONAM)
24. Macal CM, North MJ (2009) Agent-based modeling and simulation. In: Proceedings of the winter simulation conference (WSC)
25. Chunhui P, Xufang H, Shuren Z (2009) Multi-agent modeling and analysis for E-commerce transaction network based on CAS theory. In: IEEE international conference on E-business engineering 2009 (ICEBE'09)
26. Fatemeh H, Ahmad K, Mohammad DM (2011) ICMAP: an interactive tool for concept map generation to facilitate learning process. Proc Comput Sci 3:524–529

Global Structure in Social Networks with Directed Typed Edges

David B. Skillicorn and Quan Zheng

Abstract We address the problem of understanding social networks in which there are several qualitatively different edge types which capture different kinds of relationships between them. We also assume that edges may be directed since many relationships naturally have an associated direction, and properties such as influence flow asymmetrically between individuals. Our approach produces interpretable models, because we model the flow between the subgraphs of different edge types as passing only from an individual in one of her roles to another of her roles; is effective because it is based on spectral embedding techniques; and is mathematically well-motivated by analogy with lazy random walks. We illustrate the results on several real-world datasets and compare our results with the only other directed heterogeneous graph embedding technique.

1 Introduction

Social networks capture the pairwise relationships between individuals (or, in general, nodes of any kind), and induce emergent knowledge from the global structure implied by the totality of these pairwise connections. Direct algorithms for computing such emergent properties are often expensive and do not handle incremental settings well, since adding a single node to a graph can completely alter its path structure. It is common, therefore, to use spectral techniques to embed a graph in a geometric (usually Euclidean) space. Many of the desired properties can then be computed directly from the geometry.

The edges in a social network graph are allocated positive weights indicating the strength of the local similarity (or intensity of the relationships) between the

D.B. Skillicorn (✉) • Q. Zheng
School of Computing, Queen's University, Kingston, ON, Canada K7L 3N6
e-mail: skill@cs.queensu.ca; quan@cs.queensu.ca

Ş. Gündüz-Öğüdücü and A.Ş. Etaner-Uyar (eds.), *Social Networks: Analysis and Case Studies*, Lecture Notes in Social Networks, DOI 10.1007/978-3-7091-1797-2_4,

pair of nodes that each edge connects. For pairs of unconnected nodes, there is an implicit similarity derived both from the sum of the weights along each path from one to the other, and often from the number of such paths that exist. Given an embedding, these indirect similarities can be computed directly based on the reciprocal of the distances between the embedded nodes: close nodes are considered to be highly similar, even if they were not originally connected. This is the basis of *edge prediction*.

Spectral approaches to graphs, therefore, integrate local, pairwise similarity information over the entire graph, use the integrated information to embed the graph, and then determine improved local similarity information from the embedding. Even for a connected pairs of nodes, the distance between them (and so the implied similarity) may change from its original value, indicating that the global structure of the graph has implications for what at first appeared to be purely local information. The natural dimension of an embedding of a graph with n nodes is $n - 1$, so it is usual to follow an embedding by a projection to a much lower dimensional space before calculating distances. Projection to a lower dimensionality can be a kind of 'denoising' of the graph and, of course, also makes visualization possible [10].

In most social network analysis models, edges are weighted, but are of a single type. This rules out several interesting kinds of analysis. The social network of each individual consists of different kinds of relationships, for example those associated with work (colleagues) and leisure (friends) with, of course, potential overlaps. If we want to model the way in which, say, influence works in such a social network, treating all of the edges the same seems inadequate. Presumably some kinds of influence flow better to (as it were) colleagues while other kinds flow better to friends. On the other hand, it also seems inadequate to treat the colleague and friend networks as entirely separate. For example, it is plausible that influence can flow from an individual to her colleagues, and from them on to their friends [2].

The contribution of this paper is to extend conventional spectral graph analysis to networks where the edges have (one of a fixed number of) types (and positive weights). This makes it possible to combine the features of, for example, Facebook and LinkedIn into a single network framework that takes into account the qualitative differences in the functionality of edges.

As a side-effect of the construction, it is possible to answer two kinds of edge prediction questions:

1. Should there be an edge between these two nodes?
2. If there will be a new edge between two nodes, what type of edge is it likely to be?

This enables, for example, a refinement of the suggestions made by social network systems from "this is someone you might know" to "this is a potential colleague" or "this is a potential friend".

In Sect. 2, we introduce our strategy for modelling networks with multiple edge types. In Sect. 3 we explain the details of the embedding algorithm. In Sect. 4 we illustrate its application to several real-world datasets. Direct validation is, of course, not possible since there is no 'right' answer to this problem, so we justify our design choices by the explanatory power of the results. In Sect. 5, we compare our work to other approaches, most of which can only model undirected graphs. Finally, in Sect. 6 we summarize our contribution.

2 Approach

Embeddings of undirected graphs attempt to place nodes that are well-connected close together so that the (plentiful and/or heavily weighted) edges that join them are short. The implicit circularity in this description is resolved, as it often is, by the use of an eigendecomposition whose result is a geometric fixed point. An alternate, but equivalent view, is that the length of each edge in the embedding should reflect the frequency with which it is traversed in a random walk on the graph—short edges are traversed most often, and nodes with many incident edges are visited most often. Long edges reflect connections between parts of the graph that are difficult to reach from one another, and so represent good places to cut the graph if the goal is to find communities or clusters.

Direct use of an eigendecomposition of an adjacency matrix fails because well-connected nodes in the graph correspond to rows with large entries in the matrix. An eigendecomposition embeds such nodes far from the origin when, from a graph perspective, they should be central. Similarly, poorly connected nodes have rows in the adjacency matrix with small entries, so an eigendecomposition will embed them close to the origin.

Adjacency matrices are therefore transformed into one of a number of Laplacian matrices [6]—the choices embodying decisions about how edges are integrated into paths, and whether differences in local degrees should affect the distance scale in the embedding—which turn the graph structure inside-out so that well-connected points are placed centrally (and sparsely connected points are placed peripherally).

For directed graphs, the intuition behind spectral techniques is less obvious. Approaches begin with the asymmetric adjacency matrix of a directed graph and turn it, in one of a number of ways, into a symmetric matrix which can then be embedded in the standard way [3]. However, this must be done with some care if it is not to effectively ignore the directed nature of the edges. For example, in a directed graph the frequency with which a random surfer traverses an edge no longer depends only on the weighted edges incident with the nodes at both of its ends, but also on the size of the "upstream" region of its tail node. This is a global property and therefore requires an extra global computation before the embedding.

We now turn to our approach to modelling graphs with directed edges of qualitatively different types, representing relationships of different kinds that are not necessarily symmetric. We model this as a set of n nodes connected by directed edges of c different types which, for convenience, we will consider to be colors. The set of edges of each particular type (color) forms a subgraph on the set of n nodes. There could, of course, be edges of more than one type between a particular pair of nodes, and there could be nodes connected by edges of only one type; in general, nodes will have incident edges of several types. Edges are positively weighted with a value that represents the pairwise similarity between the nodes they connect and directed to indicate how properties such as influence flow.

The social-network intuition behind this model is as follows: Each individual plays a number of different roles, and in each role connects them to a set of other individuals with role-specific intensities or closeness. Properties such as influence flow along edges of a single color exactly as we assume they do in existing social-network modelling. However, paths that are made up of multiple colored edges encounter a kind of resistance that is proportional to how easily each individual can move among their roles. In other words, the cost associated with a path arises from the change from an incoming edge of one color to an outgoing edge of a different color. It is modelled as the cost of that individual "changing gear" to convey some property from one of the networks in which they participate to another. In practice, this cost might be associated with contexts: an individual has to remember something they learned at work in their home context in order to pass it on to a friend. Thus the edges of the combined network are of two kinds: 'horizontal' edges (all of a single color) within a single subgraph, and 'vertical' edges that connect the representations of the same individual in different subgraphs.

3 The Algorithm

We model a graph with n nodes and c different edge types by extracting c subgraphs, each consisting of copies of all of the nodes and the edges of a particular color. We regard each subgraph as a horizontal layer. The c copies of each node of the original graph, one in each layer, are connected by a directed clique of 'vertical' edges. Our strategy is to embed this multi-layer graph using a directed-graph spectral technique and, from this embedding, to draw conclusions about the graph's overall properties.

Each subgraph in a layer contains only the edges of a single edge type from the original graph. Thus all of the edges of the original graph appear in exactly one of the 'horizontal' subgraphs.

But what weights should be allocated to the 'vertical' edges that connect the c copies of each of the original nodes in a directed c-clique? The embedding of the entire graph depends both on the 'horizontal' pulls that affect the embedding of each layer and the orthogonal 'vertical' pulls of the vertical edges between copies—the final positions of each node in the embedding depends on the balance between these two. The vertical weights cannot therefore be chosen arbitrarily but must somehow by scaled by the strength of the horizontal similarities.

We motivate the choice of weights for the vertical edges from a random-walk perspective. One standard approach to spectral embedding begins by converting an adjacency matrix to a random-walk matrix, by dividing each row by the row sum. Such a random walk is more stable if it is made to be lazy: the probability associated with each node is divided so that half is allocated to the outgoing edges proportionally to their edge weights, and half is allocated to a self-loop at the node.

In other words, at every step there is probability 1/2 of leaving the current node via one of its edges, and probability 1/2 of remaining there for another turn. In each row of the resulting random-walk matrix, therefore, there is 1/2 on the diagonal, and the remaining entries sum to 1/2.

We take the lazy part of this approach and use it to weight the vertical edges: the total (outgoing) vertical weight associated with each colored node is the same as the sum of its total (outgoing) edge weight in its layer. This vertical edge weight is initially divided equally across the $c - 1$ edges connecting this node to its versions the other colored layers.

We therefore create a $cn \times cn$ matrix to describe the entire graph. On the main diagonal are (non-lazy) random-walk submatrices derived from the adjacency matrices of each of the layer subgraphs. The off-diagonal matrices are themselves diagonal, with entries for the vertical edges. Hence, the row sum of the main diagonal submatrix is 1, and so is the row sum of the remaining entries of any row (although we normalize this in a later step).

The algorithm uses the following six steps.

Step 1: Convert Each Layer to a Random-Walk Matrix

For simplicity, we assume that the entire graph comprising all of the typed edges is connected. However, each layer subgraph need not necessarily be connected.

In particular, there may be isolated nodes in some of the subgraphs, and this requires a special adjustment. For such an isolated node, we put a 1 on the diagonal of its row to represent a self-loop.

The adjacency matrix for each subgraph is normalized by dividing each row by its row sum to convert it to a random-walk matrix, *rw*. Therefore, the row sum of every row in a subgraph random-walk matrix is equal to one.

A further modification is needed to deal with a situation that cannot arise in the undirected graph setting. In general, a directed graph is reducible, that is it contains regions that act as sinks for random walks (once a random walker enters such a region there is no outgoing edge that leaves it). This problem may occur in any of the layer subgraph random-walk matrices. In a later step, we need to compute the principal left eigenvector of a matrix that contains these subgraph matrices, and so we must deal with the reducibility problem. We use a standard technique, also used by Google's PageRank: we define a small value ε, and convert each random-walk matrix to

$$rw_{new} = (1 - \varepsilon) * rw + \frac{\varepsilon}{n} * J$$

where J is an $n \times n$ matrix of ones. This can be thought of as allowing a random walker, at any step, to 'teleport' with probability ε from the current node to any other; or, with probability $1 - \varepsilon$ to act as before. The resulting matrices are irreducible. Intuitively, it is now not possible for a random walker to be 'trapped' in a region because there is always some small probability of moving to node outside the region. In what follows we use a value of 0.15 for ε, but the results are not very sensitive to this choice.

Step 2: Connect the Layers Together to Build a Random-Walk Matrix R

In this step, we represent the graph by a $cn \times cn$ random-walk matrix with, at each step, half the probability staying within the current layer, and half the probability being divided across the $c - 1$ outgoing vertical edges from each node.

The full matrix is made up of $n \times n$ submatrices. Those down the main diagonal represent the structure of the layer subgraphs; they are the matrices rw_{new} described in the previous step, but with their entries divided by 2, so that their row sums are all 1/2. The off-main-diagonal submatrices are $n \times n$ diagonal matrices with the value $1/(2(c - 1))$ along their diagonals. The row sum of each row of the $cn \times cn$ matrix is therefore 1.

Step 3: Compute the Importance of Each Node in Each Subgraph

We now modify the initial vertical edge weights to reflect the importance of the nodes that they connect. Intuitively, the vertical connections between a node that is important in one or more of its layers is also more important for understanding the relationships between layers. In a social networks, for example, a good strategy for finding a path to another person is to go via some well-connected intermediary. When the edges have different types, the desired intermediary is one who is well-connected in both subnetworks.

For an undirected graph, the importance of a node is simply the total incident edge weight at that node. However, for a directed graph, this is not sufficient—the importance of a node in its layer depends not only on its immediate environment (its incident edge weight) but also on how accessible its upstream environment is. For example, strong connections from A, B, and C to a node D do not automatically make D a well-connected node. If, for example, A, B, and C are only source nodes, no other node can exploit their access to reach D and D remains globally isolated. In other words, a random surfer will not spend much time at D. Conversely, a node whose local degree is low may actually be quite important because its upstream neighborhood is large, and it represents one of few exit points from that neighborhood.

The dominant left eigenvector of the $cn \times cn$ matrix integrates this information and provides a global importance vector—the larger the magnitude of the entries, the more important the corresponding node (the greater the fraction of time a random surfer spends there). We compute the absolute values of the dominant left eigenvector, and call its values π.

Step 4: Construct the Laplacian Matrix L

This matrix, L, is defined to be:

$$L = I - \frac{\Pi^{1/2} R \Pi^{-1/2} + \Pi^{-1/2} R' \Pi^{1/2}}{2}$$

where Π is a diagonal matrix whose entries are π [3]. The effect is to alter the edge weights of R based on the relative importances (the ratios of the square roots) of the nodes at each end of the edges. In particular, the vertical edge weights change to reflect the importance of the nodes at their 'tail' end—so that more influence flows vertically from nodes that are also more influential in their layer.

L is now a symmetric matrix that captures the similarity over the entire graph in a way that properly represents the underlying directed graph.

Step 5: Eigendecomposition of Laplacian Matrix L

The Laplacian matrix L can be embedded in a $cn - 1$ dimensional space using an eigendecomposition

$$L * g_i = \lambda_i * g_i$$

where the g_i are the columns of the matrix G of eigenvectors.

Step 6: Compensate for Variable Local Density

We need to alter the embedding because the eigenvectors G correspond, if this were an undirected graph, to the eigenvectors of a symmetric Laplacian. This does not represent the graph well when the nodes' total edge weights are of substantially different magnitudes, that is, the 'density' is different in different regions of the graph. Nodes with low importance values tend to be embedded closer to the origin than similar nodes with slightly greater importance, but in the natural model they should be more peripheral.

To alter the embedding to one which corresponds with a random-walk Laplacian embedding, had the graph been undirected, we multiply the eigenvectors by $\Pi^{-1/2}$:

$$f = \Pi^{-1/2} g$$

Using the entries of F, the matrix of modified eigenvectors f, produces a more plausible embedding.

The k eigenvectors corresponding to the k smallest non-zero eigenvalues give the optimal embedding of the nodes of the graph in k dimensions. Nodes are positioned so that the distance between each pair reflects, as faithfully as possible, their global similarity in the directed graph. A rendering of each of the subgraphs can be made by plotting the positions of the nodes of its color, and the edges of that color connecting them. Renderings of the relationships between subgraphs can be made by plotting subgraphs, and the 'vertical' edges between them. The lengths of a vertical edge represents how dissimilar the roles of the associated node are in the subnetworks In a social network, a person who plays qualitatively similar roles in both a social and friendship subnetwork will have a short 'vertical' edge connecting the nodes that represent their position in each subnetwork.

In the embedding, the similarity between person A and person B taking edges of all color into account is the shortest distance between their representative points of the same color. A natural question is why the distances are only calculated within single layers. What about the distances from the red version of one node to the green version of another? How does a person in one role, say social, directly influence someone else in another role, say work-related? In our approach, the only way is to influence the other person in their social role, and have it affect their work-related role via their individual personal dynamic, captured by the length of the appropriate embedded blue edge. In other words, meaningful influence paths can only include an even number of blue edges. It is possible for nodes belonging to different people and of different colors to be close. This is meaningful: it means that their view of the rest of the graph is similar, despite their different roles. For example, one person's relationships from a social perspective might be very similar to those of another person from a business perspective. This is a kind of higher-order similarity related to global positioning in a social network.

4 Examples

We cannot directly demonstrate that this embedding is better than other embeddings, because there is no ground truth for how a multi-typed social network *should* be modelled. We have argued that our approach is well motivated, and that the resulting embeddings are interpretable in reasonable ways. In particular, two copies of a monochrome graph embed as if they were a single graph, so there is no inherent distortion; each subgraph's embedding can be interpreted independently; and the lengths of edges connecting copies of the same individual's nodes show how the global differences in the subgraphs affect each individual's multiple roles.

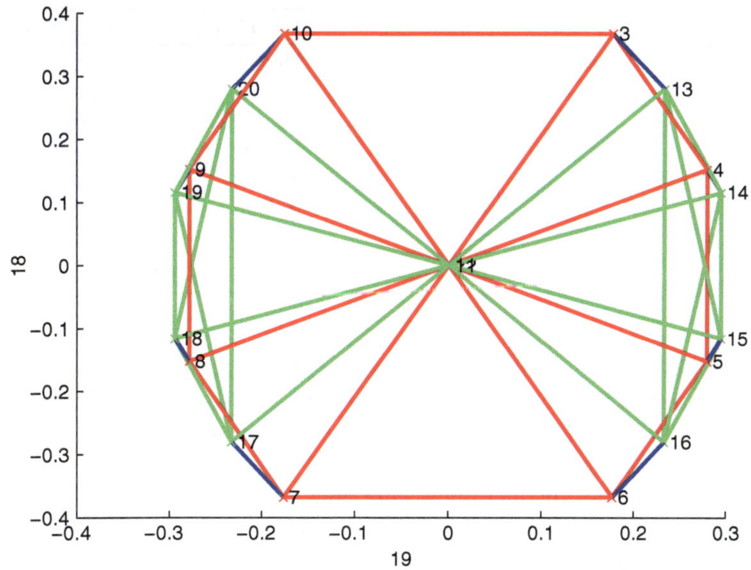

Fig. 1 The embedding of a simple graph

4.1 A Simple Example

Figure 1 shows the embedding of a graph whose red layer consists of a circle of eight nodes, all connected to two central nodes, and whose green layer consists of two cliques of size four, all of whose nodes are connected to the two central nodes. In the embedding, the distorting effect of the red layer on the green cliques pulls them into trapezoids, while the green layer pulls the red circle of points into an ellipse. The blue lines connect the two versions of each node of the graph in its embedded positions in each layer. The length of the blue edges indicates how much each node behaves differently in the domains implied by each color.

The figure shows the expected behavior: the positions in each layer are affected by those in the other layer. For example, adjacent nodes on the circumference of the circle now have differing distances because of the possibility of indirect flows via the green edges.

4.2 Florentine Families: An Undirected Graph

We now build the combined social network from two datasets describing Florentine families in the fifteenth century. This period has been extensively studied because of the rise of the Medici family, who rose from humble beginnings to dominate Tuscany and the Papacy over the ensuing two centuries.

Table 1 Division among Florentine families

Medici block	Oligarchs
Medici	Peruzzi
Tornabuoni	Castellani
Acciaiuoli	Strozzi
Ginori	Albizzi
Pazzi	Bischeri
Ridolfi	Guadagni
	Lamberteschi
Barbadori, Salviati with divided loyalties	

One popular theory explaining the growth in Medici power at the beginning of the fifteenth century is that they developed two separated social networks. On the one hand, they built financial relationships with *nouveau riche* families; on the other, they built marriage ties with the "old money", oligarch families [8]. By keeping these two kinds of ties distinct, they were able to act as gatekeepers between two communities, a role they parlayed into power broking.

The dataset describing connections among Florentine families was collected by Padgett (www.casos.cs.cmu.edu/computational_tools/datasets/sets/padgett/index.html). The undirected edges are labelled as either marriage or business ties. This dataset has been extensively analyzed from a social network perspective, but from the point of view of one of the link types at a time, or as a single network [12].

Although allegiances shifted from time to time, the families can be divided roughly into those affiliated with the Medicis, and the older, oligarch families as shown in Table 1.

Figure 2 shows the subgraph of marriage ties, embedded using two identical copies of the graph to illustrate the consistency of the method. The graph shows a division among the oligarchs between those with slightly closer ties to the Medicis (Albizzi, Lamberteschi, Guadagni) and those without (Castellani, Strozzi, Peruzzi, Bischeri).

Figure 3 shows the subgraph of financial ties. Here there is obvious structure; essentially there is a linear connection with the Medicis almost at one end, and the oligarchs at the other. The two groups of oligarchs visible in the marriage network are rearranged in the financial network so, for example, the Castellani have close financial connections to the Medici, but not close marriage connections.

The combined social network is shown in Fig. 4. Edges between the two versions of each individual node are drawn in blue. The length of these blue edges can be interpreted as an indication of how different the social and business roles of each family were in Florence at this time.

The combined graph shows that, when both kinds of relationships are taken into account, there are two different classes of families. The first include Salviati, Pazzi, Acciaiuoli, Ridolfi, and Albizzi who are all relatively close to the Medici; and the second include Guadagni, Lamberteschi, Bischeri, and Peruzzi, Strozzi, and Castellani. The combined network provides information about the network status of the two families considered to have divided loyalties. The Salviati family, on

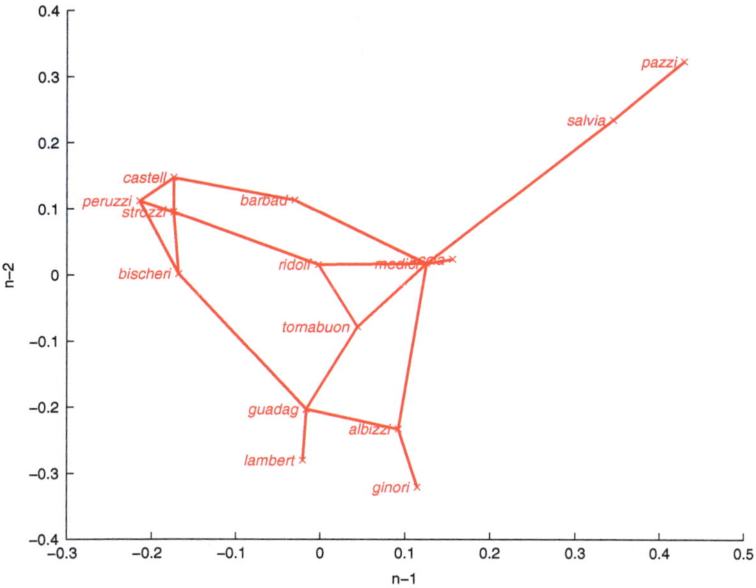

Fig. 2 Social network of marriages in the undirected data

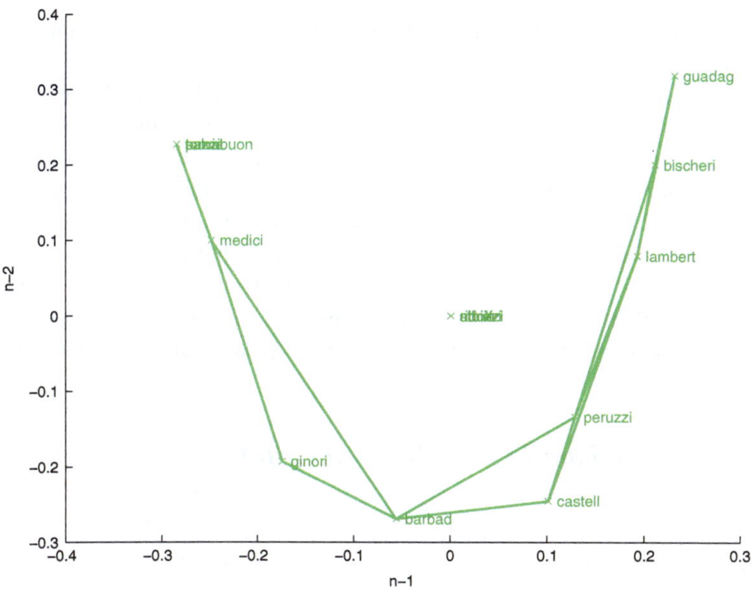

Fig. 3 Social network of financial dealings in the undirected data (nodes that are totally unconnected are embedded at the origin)

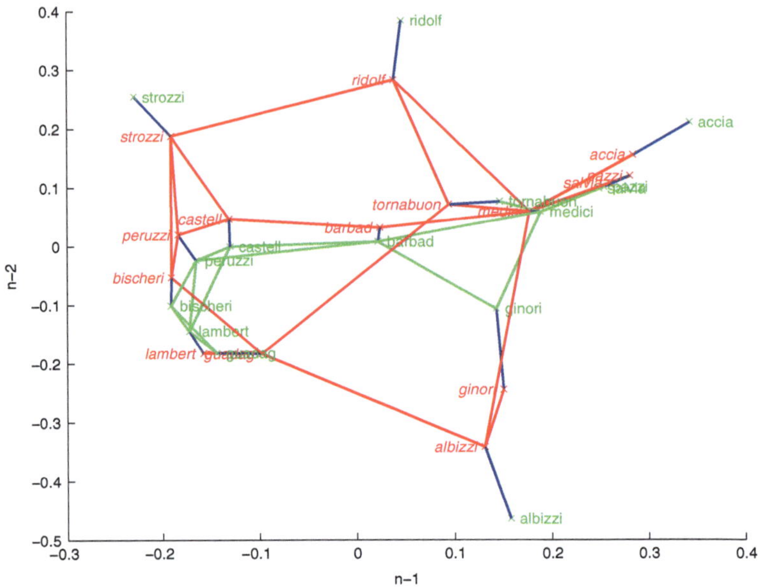

Fig. 4 Combined social network of the undirected data

this data, is clearly connected to the Medicis. The Barbadori family plays a strong intermediary role in both networks but they appear to be better associated with the Medici side.

This combined social network does not support a model in which the Medicis act as intermediaries between the newer families and the oligarchs, since the Medicis are at one extreme of this arrangement, rather than central to it. There are four different connections between the Medici group and the oligarchs: via Ridolfi, via Tornabuoni, via Ginori and Albizzi, and via Barbadori, this last being the most important (and most direct) because it involves both marriage and financial connections. The combined graph also makes it clear that the Strozzi family is the most isolated from the Medici family. Indeed, the relationship between these families became, over the next century, one of the great rivalries in Tuscany.

4.3 Florentine Families: A Directed Graph

Padgett also constructed, by hand, a larger social network [8]. This network has a much greater variety of edges, for example separating financial ties into those involving lending money, or acting as guarantor for a loan. Almost all of the edges in this network are directed.

As a second example, we analyze this larger network which requires the full power of the directed-graph embedding. We aggregate the edge types into two categories: those related to finance, and those related to personal relationships.

Table 2 Division among Florentine families

Medici block	Oligarchs	Oligarchs
Medici	Bischeri	Guadagni
Fioravanti	Ardinghelli	Da Uzzano
Dietsalvi	Altoviti	Solosmei
Davanzati	Rondinelli	Guasconi
Orlandini	Pepi	Castellani
Cocco-Donati	Peruzzi	Scambrilla
Valori	Benizzi	Strozzi
Guicciardini	Panciatichi	Aldobrandini
Ginori	Rucellai	Lamberteschi
Tornabuoni	Baroncelli	
Dall'Antella, Albizzi, Della Casa, Velluti with divided loyalties		

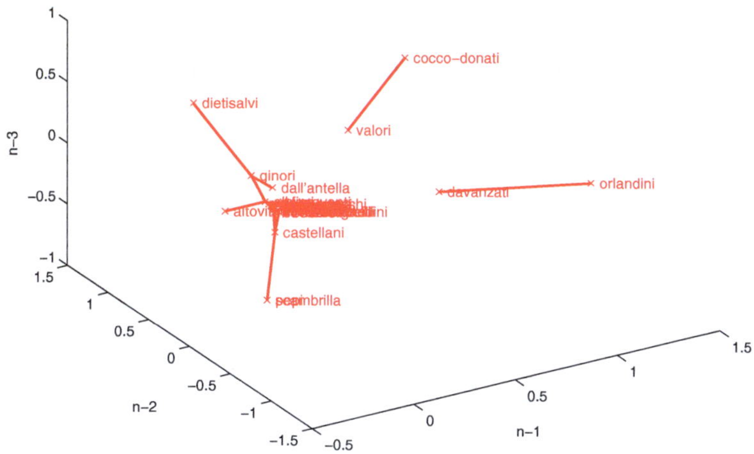

Fig. 5 Social network based on personal ties for the larger directed data (trivial edges omitted)

Financial connections include trade, bank employment, real estate, and being a guarantor; some of these edges are naturally directed and some are not. Personal relationships include personal loans, patronage, friendship, and marriage. When relationships are naturally directed, we orient them so that the more powerful actor is at the tail of the arrow and the less powerful at the head—so that properties such as influence flow along the arrow in the natural direction. Padgett orients marriage connections from the family of the wife to the family of the husband, and we preserve this—but an equally good, perhaps better, case could be made for orienting them the other way. The expanded set of families and their allegiances are shown in Table 2.

Figure 5 shows the social network based on personal ties. The network is actually completely connected via the edges added in Step 1 but we do not render these trivial edges.

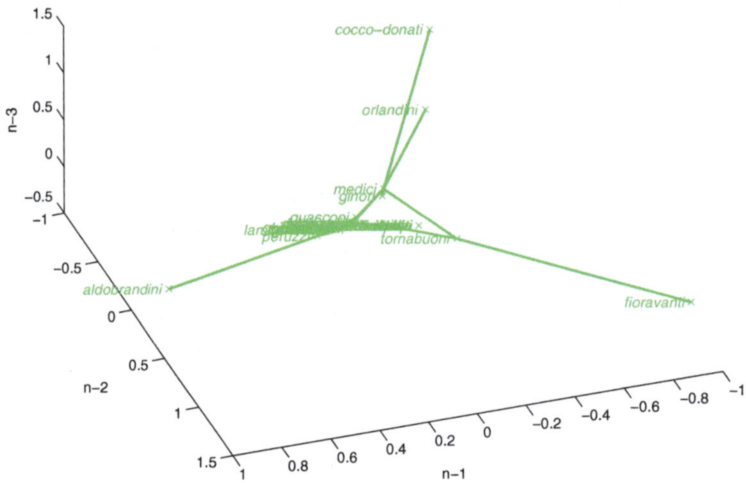

Fig. 6 Social network based on financial ties for the larger directed data

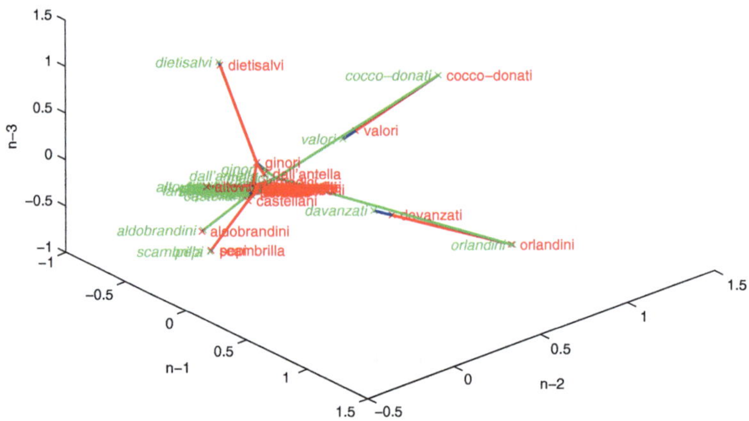

Fig. 7 Combined social network for the larger data

Figure 6 shows the social network based on financial ties.

Figure 7 shows the embedding of the combined social network. As expected, nodes with relatively poor connections are placed at the extremes, while well-connected nodes are placed centrally. While this rendering accurately reflects both the direct and indirect connections among the families, it becomes difficult to understand because of the density of nodes and edges.

Figure 8 shows the core families without the graph edges—distance is a surrogate for similarity between *all* pairs of points, there is an implicit clique connecting all of the points, and so there is no need to render the edges. This makes the figure much more intelligible.

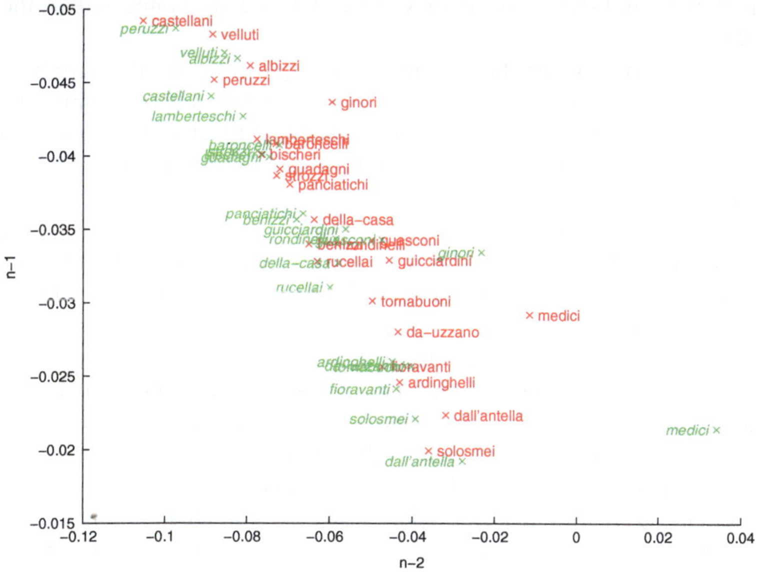

Fig. 8 The core members of the social network shown without edges

In general, most families play similar roles in both the financial and personal domains, that is their red and green points are close to one another. This is not surprising, given the close-knit nature of the Florentine community and the fact that these families would not have considered the finance and personal domains as distinct in the way that a modern audience might. There are, however, two families with large differences between their two roles: Medici, and Ginori (the financial point for Ginori is below the personal point and partly obscured).

It is also noteworthy that the structure describing these families is essentially single-factored, with families such as Albizzi at one end, then Strozzi, then Benizzi, and finally smaller families associated primarily with the Medici. For Ginori and Medici, their financial placement is higher in the figure than their personal placement; while for families such as Strozzi and Peruzzi the reverse is true. The Albizzi and Velluti families seem, from this result, to be clearly aligned with the oligarchs, at least during this early period.

Once again this data provides little evidence for a middleman role for the Medici's in Florentine society. There appears to be a central group of oligarchs, with the Medici and Ginori families as weak hangers-on; most other Medici-aligned families are much more isolated (and so not present in this view of the core).

4.4 Edge Prediction

An embedded combined social network allows us to do edge prediction because the distances between pairs of nodes in the embedding are a meaningful reflection of their global similarity, even if they were not already connected in the given subgraphs.

The first form of edge prediction answers the question: should there be an edge between these two vertices? For example, in many datasets two unconnected nodes that have a short distance between them suggest either that there is a connection that failed to be collected in the dataset, or that there should be or might soon be an edge (for example, in collaboration networks).

In the first Florentine families dataset, we can, for example, ask the question: which unconnected families are likely to be connected? In a fifteenth century version of Facebook, we could suggest to both families that the other is one with which they might build an alliance.

We do this by computing the edge lengths for each edge that is not already connected in the given datasets using the five most significant dimensions in the embedding. (As a result, the distances do not necessarily match the apparent distances in the two-dimensional view of the embedding in Fig. 4.)

The shortest distance in this embedding is the financial distance between Medici and Castellani, suggesting that Barbadori's role as middleman in financial dealings is at risk. The Medici family also has short potential marriage connections to the Peruzzi and Castellani families. The Strozzi and Medici families are quite remote from one another in this social network, but the data suggests that, if they make a connection, it is significantly more likely to be a marriage connection rather than a financial connection. And indeed that is what happened: the Strozzis married into the Medicis, but not until years later.

A second version of the edge prediction problem is this: suppose a new edge appears in the real world or because of updating the dataset. What color is it most likely to be? This question can be answered by comparing the distances between the relevant pairs of nodes in the red layer and in the green layer, and predicting the edge type will be whichever of these distances is the smaller. For example, the Bischeri and Castellani family are already fairly well connected by both marriage and financial links, but do not have a direct connection. If we learn that these two families have made a direct alliance, which mechanism is more likely to have been used? The marriage distance between them is 0.109 and the financial distance is 0.114 so we predict that a marriage relationship is slightly more likely.

5 Related Work

In general, there have been two different ways of framing the problem of networks with heterogeneous or typed edges, although the resulting algorithms often have some commonalities. In the first, the subgraphs are regarded as different (perhaps partial) views of the same underlying graph. This is particularly common in applications involving images, where different perspectives can produce views that appear different but are actually describing 'the same thing'. In the second, the subgraphs are regarded as inherently different, and the goal is to find a consensus or average graph that adequately represents the subgraph structures (by alignment, for example). Our approach differs from both of these because we maintain the subgraphs in their entirety in an embedding that represents them in a mutually consistent way. Any desired alignments, commonalities, or averaging can still be carried out, but as a subsequent step.

Technically, algorithms can be roughly separated into those that embed the subgraphs first, and then attempt to 'paste' these spaces together into a consistent whole; and those that construct a composite structure and then embed this composite directly.

As an example of the first kind, Kumar and Daumé [5] use iterative techniques, using each representation of the graph in turn to converge to a common representation. Cheng and Zhao [1] combine the distances in each separate embedding to create a single similarity matrix and then repeat the spectral clustering on this matrix to produce the final embedding. Tang et al. [11] use an approach they call Linked Matrix Factorization to decompose each graph and then combine pieces of the decompositions with weights to create a global fusion of the data. The problem with these strategies is deciding how the individual representations should be combined—there is not usually enough information to decide this in a principled way.

The second kind of algorithm connects the subgraphs representing each view of the data with edges whose weights are then determined using a regularization framework [4, 7]. Xia et al. [13] make a similar assumption to Kumar and Daumé that there is an underlying correct embedding (because they are interested in images). They use an alternating optimization algorithm to estimate a common embedding and weight each subgraph by its similarity to this common embedding. De Sa [9] considers the two subgraph case, models it as a bipartite graph and uses a spectral partition of the product matrices from the graph. Regularization is problematic because it is not clear whether to build a global regularizer that applies equally to edges of each type *and* to the edges that connect different subgraphs; or whether each subgraph should have its own regularizer and there should also be another regularizer for the connecting edges. There does not seem to be a compelling argument for either possibility, let along a way to relate them to one another.

All of this related work applies only to undirected graphs. The closest work to ours is by Zhou and Burges [14]. They construct random-walk matrices for each subgraph, but they then create a mixture of these random-walk matrices, use the Chung-style symmetrization [3] as we do, and embed this (combined) matrix. Their strategy therefore finds an embedding that reflect each of the subgraphs reasonably well, provided that the subgraphs are not too different.

Figure 9 shows the embedding of the second dataset using the Zhou and Burges algorithm. It resembles the embedding we obtain but loses available information because of the averaging. This is shown more clearly in the analogue of Fig. 8 shown in Fig. 10. The same single-factored structure is evident, but the distribution of points along it is more uniform and, of course, there is no information about the difference between financial and personal roles.

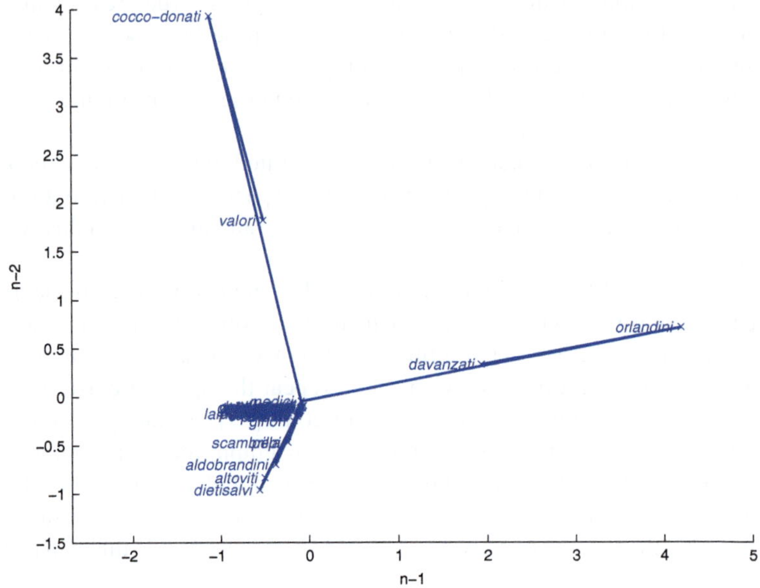

Fig. 9 The Zhou–Burges embedding of the larger dataset

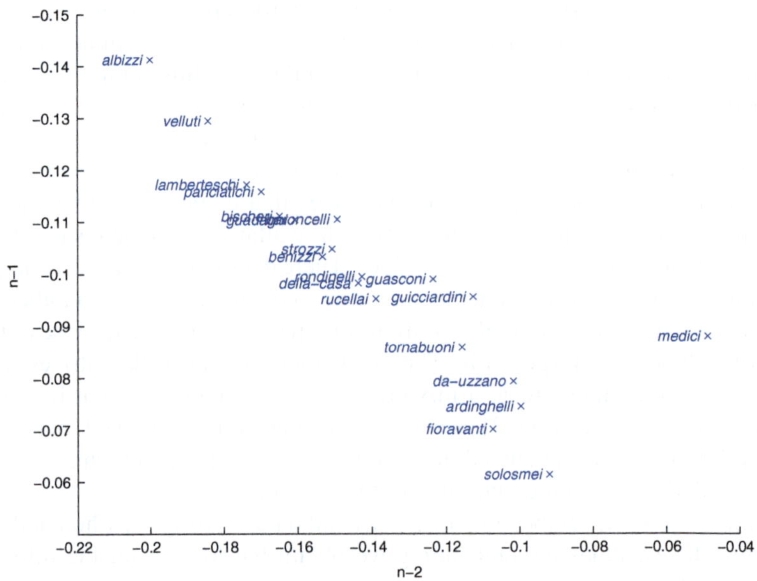

Fig. 10 The core region using the Zhou–Burges embedding of the larger dataset

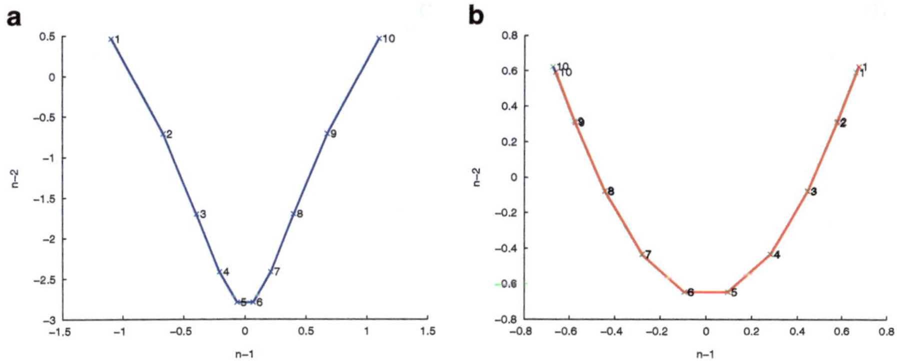

Fig. 11 Comparison of embeddings of a path directed in opposite directions in each subgraph. (**a**) Zhou–Burges embedding; (**b**) our embedding

The Zhou–Burges algorithm does poorly when the structure represented by the subgraphs is substantially different. As an extreme case, consider a path of length 10, directed one way in one subgraph and the other way in the other. The resulting embeddings are shown in Fig. 11.

In the first subgraph, Node 1 has very little importance so its outgoing edge ends up with a low edge weight, and there is no incoming edge the other way. In the second subgraph, Node 1 has much importance so its incoming edge is heavily weighted but it has no outgoing edges. After the (admittedly complex) averaging that happens in the preparation and symmetrization of the Zhou–Burges algorithm, Node 1 appears to be weakly connected to the rest. In contrast, in our algorithm the importance calculation is done for the entire graph including vertical edges and this results in a much more homogeneous allocation of importance, and so a much more even spacing in the embedding. The nodes at the ends show the greatest separation between the two matching nodes as expected, because one color has high importance and the other low, but the effect is small.

Another example illustrating the differences between the Zhou and Burges algorithm and ours is shown in Fig. 12. The graph consists of a directed path from nodes 1–2–3–4, and a short-circuiting directed path from node 2 via the remaining nodes ending at node 3.

Although the embeddings seem quite similar, there are two differences. First, our algorithm makes it clear that nodes 1 and 2 are very similar and this is less clear in the Zhou–Burges embedding. Second, in our embedding the length of the embedded edge from 2 to 3 remains almost exactly the same length as the length of the loop grows, but its embedded length continues to grow in the Zhou–Burges embedding (not shown). In other words, the implicit averaging causes, for example, nodes 1 and 4 to seem further and further apart as the size of the graph (and so of the loop) grows. In contrast, our algorithm uses the presence of the green edge to keep the embedded distance between nodes 1 and 4 almost constant.

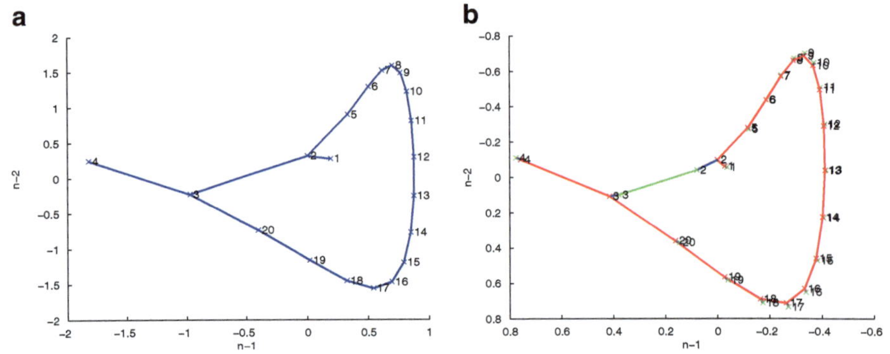

Fig. 12 Comparison of embeddings of a long directed path, shortcircuited by an edge of a different type. (**a**) Zhou–Burges embedding; (**b**) our embedding

6 Discussion and Conclusions

Social networks that ignore the different possible kinds of relationships, as most do today, surely miss useful information about how properties such as influence work in heterogeneous networks. The flows along friendship and colleague edges are different because the contexts in which each takes place are usually different, geographically and mentally, and exacerbated by the amounts of time we spend in these different contexts. On the other hand, treating, say, friendship and business networks as distinct (Facebook and LinkedIn) also surely misses parts of the interactions among them.

We have developed a spectral approach to understanding and analyzing graphs with a finite number of edge types that provides well-motivated results, and a deeper intuition to help understand the resulting embedding. There is always a problem of validation in such approaches because ground truth is not well-defined; so we argue from practical examples that the promised understanding and results are consistent with the data. The edge-typed data provides a more nuanced understanding of how these real-world social networks are structured (and incidentally suggesting that a popular model of how Medici power was deployed through the fifteenth century is not supported by this data, at least during this (early) period of Florentine history).

Our use of a directed graph algorithm enables the approach to be generalized to more-realistic scenarios in which the individual typed edges are themselves directed. In the Florentine families, for example, marriage has very different implications from the point of view of the man's family than from the woman's; and a business relationship is very different from the point of view of the borrower than from that of the lender. These effects can be accounted for directly in the framework we have constructed.

The nominal size of the matrices describing the networks grows quadratically with the number of edge types, but the extra pieces are only diagonal matrices, so sparse matrix representations keep the actual growth linear. The embedding algorithms therefore scale well because sparse matrix techniques can be used, and only a few eigenvectors are needed.

References

1. Cheng Y, Zhao R (2009) Multiview spectral clustering via ensemble. In: IEEE International Conference on Granular Computing, 2009 (GRC '09), pp 101–106
2. Christakis N, Fowler J (2013) Social contagion theory: examining dynamic social networks and human behavior. Stat Med 32(4):556–577
3. Chung F (2005) Laplacians and the Cheeger inequality for directed graphs. Ann Comb 9:1–19
4. Dong X, Frossard P, Vandergheynst P, Nefedov N (2011) Clustering with multi-layer graphs: a spectral perspective. Technical report, Computing Research Repository
5. Kumar A, Daumé H (2011) A co-training approach for multi-view spectral clustering. Computer 94(5):393–400
6. Luxburg U (2007) A tutorial on spectral clustering. Stat Comput 17(4):395–416
7. Muthukrishnan P, Radev DR, Mei Q (2010) Edge weight regularization over multiple graphs for similarity learning. In: Webb GI, Liu B, Zhang C, Gunopulos D, Wu X (eds) IEEE international conference on data mining. IEEE Computer Society, Silver Spring, pp 374–383
8. Padgett J, Ansell C (1993) Robust action and the rise of the Medici. Am J Sociol 96(6): 1259–1319
9. de Sa VR (2005) Spectral clustering with two views. In: Proceedings of the workshop on learning with multiple views, 22nd ICML. ICML (international conference on machine learning)
10. Skillicorn D (2007) Understanding complex datasets: data mining with matrix decompositions. CRC Press, Boca Raton
11. Tang W, Lu Z, Dhillon IS (2009) Clustering with multiple graphs. In: Wang W, Kargupta H, Ranka S, Yu P, Wu X (eds) IEEE international conference on data mining. IEEE Computer Society, Silver Spring, pp 1016–1021
12. Wasserman S, Faust K (1994) Social network analysis: methods and applications. Cambridge University Press, Cambridge
13. Xia T, Tao D, Mei T, Zhang Y (2010) Multiview spectral embedding. IEEE Trans Syst Man Cybern Part B 40(6):1438–1446
14. Zhou D, Burges C (2007) Spectral clustering and transductive learning with multiple views. In: Proceedings of the 24th international conference on machine learning (ICML), pp 1159–1166

Social Networks and Group Effectiveness: The Role of External Network Ties

Fabiola Bertolotti and Maria Rita Tagliaventi

Abstract This research explores the relationship between group effectiveness and social networks. Through a 5-month ethnographic observation within three work groups employed in one of the major Italian fashion firm, we recorded all interactions occurring within the groups and outside the groups' boundaries, thereby deriving the enacted communication network. Then, by means of structured interviews, we collected evaluations of group effectiveness. The evaluations given to the three groups differ and such difference cannot be traced back to the amount of communication network activated nor to the level of group members' competencies, nor to their internal network structure. The field evidence suggests that the better evaluation received by one group relates to the quality of the relationships it sets up with external actors. This group assumes a coordinating role in the whole product development process: in particular, it spontaneously triggers, through reciprocal interactions, modalities of collaborative design, not formally required, which are rewarded by organizational members' higher evaluations. The study has implications for social networks research by pointing to the importance to grasp the actual content of network relationships, thus going beyond the assessment of their presence and/or strength, in order to fully comprehend how network ties really influence organizational members' perceptions and actions.

F. Bertolotti (✉)
DISMI, University of Modena and Reggio Emilia, Reggio Emilia, Italy
e-mail: fabiola.bertolotti@unimore.it

M.R. Tagliaventi
Department of Management, University of Bologna, Bologna, Italy
e-mail: maria.tagliaventi@unibo.it

Ş. Gündüz-Öğüdücü and A.Ş. Etaner-Uyar (eds.), *Social Networks: Analysis and Case Studies*, Lecture Notes in Social Networks, DOI 10.1007/978-3-7091-1797-2_5,

1 Introduction

Organizations intensively employ work teams to perform knowledge intensive and creative tasks, and coordinate their activities[1–3]. As a consequence, numerous studies have examined the factors and processes that relate positively to the various criteria of team effectiveness [2]. Teams' effectiveness has been associated, for instance, to composition variables [4]; task characteristics and level of interdependence among members [5, 6]; leadership style [7–9]; internal states such as cohesion (e.g., [10]), shared mental models [11], cross-understanding [12], climate [13], and the presence of knowledge integration capabilities [14].

Recently, research on teams has started to pay growing attention to the network perspective [15] focusing on the ongoing social relationships in which organizational actors and teams are embedded. The research streams that concentrate on the dynamics related, respectively, to within-team networks and to between-team social networks can put forward comprehensive insights of social networks in team contexts, and can help us appreciate how teams work within the larger organizational contexts and what outcomes they generate (e.g. [16–21]). However, while at an individual level of analysis the positive relation between centrality in a network and performance has been well acknowledged [16, 22–24], at a group level of analysis empirical evidence is still controversial. Seminal studies focusing on boundaries management support the idea that work groups interpreting external activities—such as diagnosis of clients needs, feedback seeking, and selling services—as their primary goal are rewarded with supervisory higher appraisal [25, 26]. Later, the results of Baldwin et al. [24] and Sparrowe et al. [16] have questioned the existence of a strong relationship between work groups' effectiveness and the characteristics of their internal and external communication networks. Other contributions problematize such issues suggesting that a large number of external ties could be either detrimental or favorable to group performance based on several factors like, for instance, autonomy and task overload [27].

The purpose of this study is to extend this stream of research in two ways. On one hand, we offer new insights on the literature on group effectiveness by bringing additional evidence on the relationship between groups' communication network structures and groups' effectiveness. On the other hand, we contribute to the network literature by supporting the importance to focus not only on the intensity and flow of relationships within a communication network (collected via self-reported data), but also on the relationship actual content as gathered through participant observations within groups.

This chapter is divided into several sections. In the first part we offer an overview of the theoretical background on group effectiveness and social networks, underlying research questions that need further exploration. We analyzed three groups working in a major Italian clothing company. We collected data on the enacted network structure through participant observations, and we collected performance information through structured interviews. Then we discuss the evidence from the field. Conclusions, limitations of the study, and a future research agenda constitute the last section of this chapter.

2 Theoretical Background on Team Effectiveness and Social Networks

Because of the widespread use of teams in organizations, scholars and practitioners have devoted considerable attention to exploring the predictors of team effectiveness in terms of both design factors, e.g., size, composition, diversity, and emerging social processes, e.g., conflict, trust, identification (for an extensive review see [2]). Hackman [28], in his seminal contribution dating back to 1987 and subsequent work [29], provides a definition that has become widely accepted. He defines effectiveness as being partially about performance (teams must produce outputs considered as adequate by those who receive, or who are in charge of evaluating, them), but also about the ability of group members to work together in successive tasks, and the satisfaction that group members derive from the execution of their tasks. Relatedly, Mathieu et al. [2], when explaining the Input-Mediator-Outcome (IMO) team effectiveness framework, advocate for the relevance of members' affect and viability in addition to performance measures based on team outcomes and roles only.

Current theories highlight the importance of informal social interactions for exchanging information and knowledge, and enhancing collaboration [30]. Consequently, group relations with external parties, as well as the patterns of relations (both formal and informal) among group members, have been related to work teams' performance lately [16–20, 25, 26, 31–33]. Understanding the social network in a team context, i.e., the social relations at a group level of analysis, is important because teams are inclined to enact these types of relationships within and outside their boundaries [24].

Baldwin et al. [24] analyzed the impact of centrality in friendship and communication networks of MBA team members upon the performance of the team, but found no support for the hypothesis that individual centrality indexes were positively related to team level performance. Likewise, communication between different teams was not significantly related to team performance, while within-team communication was. Sparrowe et al. [16] extended these previous studies by delving into both the relationship between individual network structure and individual performance, and the relationship between group performance and two structural characteristics of the informal advice network (namely, group density, and level of centralization). They hypothesize a positive relation between group density and performance because numerous mutual interactions increase interdependence among members, which, in turn, enhances cooperation and the subsequent performance. On the contrary, the authors posit that group centralization produces lower levels of group performance because the scarce involvement of all group members in the advice network reduces those interdependences and cooperation that are vital for the completion of the work. Their results support the hypothesis that individual centrality in a group network is positively related to individual performance, whereas group density and centralization are not significantly related to group performance as evaluated by supervisors. Conversely, Janhonen and Johanson [33] account for a positive relationship between team communication network density and performance. Such evidence is consistent with the meta-analysis of studies investigating the relationship between team performance and within-team network structure performed by Balkundi and Harrison [20], who found that team viability and task performance increase in teams characterized by more dense interpersonal ties. Also Reagans and Zuckerman [17] testify to the positive consequences of frequent communication among members of corporate R&D teams (that is to say, high communication network density) for their performance. In addition, the authors also found that network heterogeneity, expressed as the extent to which group members spend a significant proportion of time interacting with team members having diverse organizational tenure, relates positively to team performance. These latter interactions supposedly bring knowledge from outside members into teams, thus boosting teams' productivity. Mehra et al. [34], when investigating the relationship between friendship network's structural characteristics and groups' performance in 88 sales groups, found mixed results, however: friendship network density was positively related to one dimension of performance, i.e., customer loyalty, but not to sales amount.

Finally, in a recent contribution, Grund [35] confirms previous findings about the positive effect of within-team density and the negative effect of centralization on team performance through an analysis of longitudinal panel-data of 23 soccer teams, thus tackling the issue of causality between team network structure and performance.

Other research has devoted attention to the study of the type and amount of activities that work groups initiate with their environment to deal with external demands. Major contributions to this subject can be originally traced back to the empirical studies by Ancona [25] and Ancona and Caldwell [26] on the so-called 'external perspective'. The basic assumption of this perspective is that work groups do not function as close systems, but they influence, and are influenced by, their environment. Consequently, different groups' outcomes and internal processes stem from different approaches in handling external demands. Through a qualitative study conducted on five cross-functional groups, Ancona [25] explores the extent to which different interaction strategies, activated by the groups toward their clients, explain the variance in the performance evaluation expressed by group supervisors. Evidence collected during the first 5 months of life of these groups shows that low performing teams were the ones that set internal communications and a positive climate as their first priority, while limiting their relations with external actors. Conversely, the best performing groups engaged in two-way communications with the external environment and showed continuous efforts to diagnose clients' needs, seek feedback, and increase selling services. In addition, these latter groups were led by directive leaders who did not consider internal dynamics as the groups' first priority. Ancona and Caldwell [26] enriched the external perspective by claiming that it is not only the intensity of the flow, but mainly the content of the activities played out by groups toward their environment that makes a difference in terms of group performance. The groups examined, while somewhat similar in terms of the frequency of their interactions with the environment, performed very differently due to the different types of boundary management activities they engaged in. In particular, the authors found that groups activating a large amount of 'ambassador activities' (aimed at protecting the group from outside pressure, persuading external actors to support the team, and lobbying for resources) or 'task coordinator activities' (aimed at coordinating specific technical tasks and seeking feedback on their development) gained better performance evaluations than groups involved in a large amount of 'scouting activities' (aimed at obtaining information and ideas about the competitors, the market and the technology) or groups that were mainly isolated. They studied 45 new product development teams: once again they were temporary groups, in charge of carrying out new and highly uncertain tasks. Wong [32] has lately corroborated the assumption that external ties are conduits of heterogeneous and novel knowledge by positing that a wide range of external advice network increases within-group task variety, thereby enhancing group effectiveness. Research on top management teams has been particularly interesting in tapping into the relevance of external ties for team and firm performance (e.g., [36]). Studies on this kind of teams have advocated for the positive effect that top managers' relationships with members of other organizations within and outside the firm's industry can exert on top management ability to foresee and interpret environmental changes and elaborate successful strategies [37, 38].

Recently, some scholars have advocated the need to investigate team internal and external social networks as not independent from each other, but to understand how they together can predict teams' performance. Shan et al. [39] examine the friendship network: they advance the interesting idea that, in order to positively influence teams performance, teams' internal and external networks should 'match' each other. More in detail, teams with strong internal friendship networks benefit from strong external networks because the internal ties enable team members to reflect and incorporate what they have learned from external parties. Conversely, teams characterized by weak internal networks of friendship would not be able to offset the costs and efforts that accessing external parties implies. A recent contribution by Chung and Jackson [40] links the effect of internal and external ties on team performance to task routines. Since performing non-routine tasks may require team members to process complex information, teams may benefit from work relationships that cut across the team boundaries. They argue that teams performing routine tasks can maximize their team performance by reducing internal work networking and external information networking as much as possible. On the contrary, teams performing non-routine tasks can maximize their team performance by increasing internal work and trust networking relationships up to a moderate level, following an inverted U-shaped relationship, and enhancing external informational networking as much as possible.

We aim at extending this stream of research by examining the impact of external network ties upon group effectiveness. As opposed to previous research, the groups that we analyze are intra-functional and stable over time, both in terms of team membership and life span. More specifically, we intend to explore if, in the specific setting of work groups that are part of a manufacturing process, but largely independent from other organizational units for resources allocation, centrality in the organizational social networks acts as a meaningful variable in predicting group performance evaluation. We do so by analyzing not only the impact of the quantity of relations enacted by groups, but also their content and quality. To map the group networks in terms of extant relationships and their contents, we collected data mainly through a field study based on long-term participant observations within three work groups. This choice allowed us to catch the social networks actually enacted by the groups, whereas the studies cited above grasped the social networks through group members' self-reported data.

3 Methods

3.1 The Research Site

We collected data both through qualitative techniques (observations) and quantitative techniques (structured interviews) within three groups operating in a major Italian clothing company located in central Italy.

The phases to develop a fashion collection are: prototyping (when the ideas of fashion designers turn into prototypes), sales campaign, production, and delivery of the clothes to the market. The whole cycle, which lasts approximately 6 months, is marked by the core deadline of the week-long fashion weeks held in Milan for the presentation of the new collection (one in the autumn/winter season and one in the spring/summer season). Being the fashion weeks' dates strictly given, prototyping activities are hectic and oriented towards meeting those deadlines.

The groups observed, technically called 'Collection development groups', act as gatekeepers between the fashion firm and the fashion designers (usually external to the firm). Fashion designers, who are only in charge of the creative aspects of the collection, do not participate in sales and production phases. Collection development groups are pivotal in fashion firms because they play a key role in the industrialization of fashion collections. They receive the sketches from the fashion designers, need to understand and interpret the 'imprint' that fashion designers want to confer to the specific fashion collections, and search for all the materials (fabrics and accessories) necessary to carry out the different patterns. At the same time, they 'pass' the sketches on to the pattern-making groups in charge of creating the fabric prototype, thus providing technical and style support for these other groups. They also supervise the activities necessary to complete the prototyping phase, and schedule the official dates for prototype fitting together with the fashion designer and the pattern making groups' supervisors.

The three groups that we have studied work for three different fashion designers respectively and, in order to ensure confidentiality to our informers, we will name the groups as J, M and A. The three groups show the same formal structure and are composed by a group supervisor and three group members.[1] Stable, well-defined, and full-time membership characterizes these groups, even though their products change from collection to collection. Major uncertainties are due to the fashion designers' continuous requests for change, and to the suppliers' ability to respect delivery time, which jointly make it complicated to respect the fashion week schedule and render the use of informal collaboration, above and beyond established roles and procedures, essential.

The 12 actors are all female, the average age is 31.4 years (S.D. = 5.6), the average tenure in the company is 5.8 years (S.D. = 3.6), while the average number of years spent in the collection development groups is 3.8 years (S.D. = 3.2). As for educational attainment, 11 group members hold a high school diploma and one has a bachelor degree in Arts. Members of each group share the same office that is laid out as an open space.

3.2 *Data Collection*

During the first phase of the research, we conducted observations within the three collection development groups for the period of time available for the creation of a whole fashion collection.[2] The groups' working hours were 8.30 a.m. to 12.30 p.m., and 2.30 p.m. to 6.30 p.m., from Monday to Friday. We planned observations to ensure appropriate time sampling: we spread them evenly over the 8 daily working hours and the 5 weekly working days. We each spent, on average, 5 h per day making observations, 3 days per week. The total number of observation hours amounts to 200 for J, 210 for M and 190 for A. During the observations, we took notes on the activities performed within the group, as well as on the interactions taking place among group members, and between group members and actors external to the group or to the organization, while respecting the actors' original language. Sometimes, we took coffee breaks or a quick lunch with our informants. Additionally, we were invited to take part in company meetings. At the end of each day we transcribed field notes into files.

Field notes represented the basis for coding, i.e., for tracing back different phenomena to main categories. According to the guidelines suggested by Strauss and Corbin [41], we constantly reviewed the field notes and let new ideas emerge from them, following a process of continuous intertwining between data collection and data analysis, in an attempt to discover recurrent classes of phenomena in the data [42, p. 126].

---

[1]Occasionally, three trainees worked in the three groups throughout the observation time length.

[2]Group members were made aware of our identity as researchers interested in the study of the dynamics of work groups. From the beginning, we informed them that we would take notes on what was going on within the group. To reduce intrusion we worked with small note pads. At the end of the study, we briefed our informants about the main findings of our research.

We began the analysis of the field data with a general framework in mind: to disclose the relational profiles of the three groups to understand if differences in the groups' communication networks led to differences in their effectiveness. Thus, the interactions taking place within the groups and between group members and actors external to the groups became our core category. Accordingly, we detailed properties and dimensions of interactions as follows: date, observer, active actor (who triggered the interaction), passive actor (the recipient of the interaction), communication means (e.g. face to face, telephone, or email), type of interdependence, and drivers. According to the type of interdependence, we classified interactions as sequential (specifying an input–output sequence of activities) or reciprocal (where one actor's outputs become another actor's inputs, and vice versa) interdependencies. As for the drivers, an initial briefing of the field notes highlighted that, while interactions appeared to be predominantly task-related, nonetheless social and emotional interactions were not marginal. We first coded every interaction as task-related (instrumental to the achievement of the group outcome, which is, in this case, the completion of all the prototypes) or social (instrumental to the life of the group, but not directly connected to the final outcome). We further identified the following drivers for each interaction:

- Information request
- Orientation request
- Advice request
- Information communication
- Orientation communication
- Advise communication
- Reporting problems
- Reporting solution to a problem
- Material requested
- Material provided
- Work assignment
- Taking charge of responsibilities
- Showing satisfaction/agreement
- Showing dissatisfaction/disagreement

As for the driver, it is worth underlining that many dimensions emerging from our coding are in common with coding systems already known in the literature. For instance, the first six dimensions represent the categories that Bales [43, 44] defines as task-oriented. The last two dimensions relate to interactions that affect the socio-emotional sphere. Unlike Bales' view, the evidence from the field made us believe that the same drivers of interactions apply both to interactions aimed at task completion and to interactions bounded to the socio-emotional area.

To guarantee homogeneity of coding criteria, each researcher, after independently coding her own field notes, systematically coded a randomly chosen 30 % of field notes recorded by the other researcher and compared the coding criteria at weekly meetings. We reconciled disagreements through discussion.

The coding applied to interactions enabled us to obtain a set of actor-by-actor adjacency matrices, in which the cell a_{ij} reports the number of interactions activated by the actor i that involve the actor j, in relation to different properties of interactions. We applied social network analysis techniques to matrices [45, 46]. For each matrix, we calculated the actors' centrality in terms of their outdegree and indegree [46]. Outdegree represents the number of times an actor initiated an interaction, while indegree measures the number of times an actor was drawn into an interaction by others. Degrees allowed us to determine which actors were central in the group and which were peripheral [46–48]. Coding and analyzing interactions allowed us to derive the structure of the communication networks enacted by the three collection development groups. Accordingly, the system of roles and relations among roles emerged as the recurrent schemes of interaction processes between actors [49].

In a more specific attempt to analyze the relationship between group effectiveness and social networks, we integrated qualitative techniques with a quantitative technique such as structured interviews administered at the end of the observation period. While observations allowed us to grasp the enacted communication networks, the interviews provided us with information about effectiveness measures.

At the end of the observation period we conducted structured interviews with 57 actors including the collection development groups members, top management, and first and second-level managers belonging to all the organizational units. Interviews lasted between 40 and 100 min according to the interviewees' degree of involvement. In the first part of the interviews we collected demographic data. In the second section, which we describe in detail in the next paragraphs, we grasped the group effectiveness evaluations.

3.3 Work Group Effectiveness

Regarding the appraisal of groups effectiveness, it is worth noting that the firm under study does not have a formal system of performance appraisal in place, neither for individuals, nor for groups or organizational units. Therefore we availed ourselves of the interviews to collect three measures of group effectiveness:

Group members' evaluation of effectiveness. Members of the three groups rated the effectiveness of their own group along seven items including[3]: quality of work done, completing work on time, timeliness on problem solving, ability to provide innovative products, coping with uncertainty, fashion designer' satisfaction, overall performance. We employed a seven-point scale ranging from 'very poor' (1) to 'outstanding' (7).

---

[3]Five items were drawn from Campion et al. [50] and Sparrowe et al. [16], while the remaining two were specifically developed for this study.

Senior managers' and peers' evaluation of effectiveness. Consistent with the idea that effective groups produce outputs considered as adequate by those who receive or who are in charge of evaluating them, top management and all the middle managers and supervisors in different functional departments rated the collection development groups effectiveness on ten items, including the above seven items and three additional items relevant to the management view of group effectiveness: quality of the relations among group members, cooperation with other departments, and taking charge (ability to generate new ideas). The same seven-point scale was used.

Group members' satisfaction. The questionnaire included 13 items on different topics such as satisfaction with work, colleagues and supervisors, satisfaction with wage, opportunity for growth, autonomy, trust in management. Again we used a seven-point scale ranging from 'very dissatisfied' (1), 'neither satisfied nor dissatisfied' (4), 'very satisfied' (7).

3.4 Group Composition Variables

The tasks carried out by the collection development groups do not require a specific background. For this reason, we assumed that group tenure would act as a good proxy for the ability to successfully carry out the work. Moreover, we used, as an additional measure of group members' ability to succeed in their work, the evaluations that each group member gave to herself and to group mates in relation to six core competences: autonomy, flexibility, ability to cope with stress, relational capability, negotiation ability, and product knowledge. We derived the above competencies from pilot interviews conducted with four fashion experts employed in four competing fashion companies. Experts indicated ten competencies that they believed represented core competencies for people working in collection development offices: we chose the six ones that were quoted by all the four experts.

4 Evidence from the Field

In Table 1, we provide descriptive statistics of the effectiveness evaluations that the three collection development groups received by senior managers and peers. Differences in the average evaluation assigned to the three groups emerged from the survey and, specifically, the evaluations of group J are notably higher than those of groups A and M.

Differences in evaluations are not attributable to differences in group members' abilities that would make them more or less effective in performing their job. In fact, group J's members, with the only exception of the group leader, received the lowest rates in individual competencies as shown in Table 2.

Table 1 Effectiveness evaluations of the three groups by top managers and peers

	Group J	Group M	Group A
Mean	5.3	4.7	4.6
Standard deviation	0.8	1.1	0.9

Table 2 Competences of members of the collection development groups

	1	2	3	4	5	6	Mean	Std dev
Group J							4.9	0.5
JL	6.3	6.3	6.7	6	7	6.7	6.5	0.4
J1	5.7	5.7	5.7	4	4.7	5.7	5.2	0.7
J2	4	4	3.7	3	3.7	4	3.7	0.4
J3	4	4.7	4.7	3.3	3.7	4.3	4.1	0.5
Group M							5.3	0.7
ML	4	6	5.5	5.5	6	5.3	5.4	0.7
M1	5.5	6.3	6	4.5	6	5.5	5.6	0.6
M2	5.5	5.5	5.5	3	5	4.2	4.8	1.0
M3	5.5	4.5	5.8	6.3	5.3	5.3	5.4	0.6
Group A							5.0	0.7
AL	4.5	5	6.5	6	6.5	5.5	5.7	0.8
A1	4.8	5	4.5	4.8	5	5.3	4.9	0.3
A2	4.8	5.8	6	4.5	4.8	6	5.3	0.7
A3	3.5	4	4.8	2.8	4.5	5.3	4.1	0.9

Our subsequent analysis aimed to understand if networks variables were better predictors of differences in the evaluations of groups' effectiveness given by peers and senior managers. We derived the three groups' relational profiles from the large amount of interactions recorded during our presence in the field. Out of 9.172 interactions recorded, 23.11 % (equal to 2.119) involved the top performer group J, 45.42 % (equal to 4.165) comprised M's group members, and 31.47 % (corresponding to 2.886) involved group A. The analysis of the communication networks enacted within groups highlights that groups J and M, respectively the best and the worst performer, show similar patterns of interactions in terms of density and centralization, while group A displays a pretty different internal network structure. Table 3 reports degree centrality measures for task-oriented interactions and social interactions in the three groups.

Concerning task-related interactions, groups J and M are highly centralized in terms of their outdegree: leaders, as compared to team members, are the most active actors in initiating task-related interactions. Indegree values, conversely, suggest that all members of the teams are drawn into a comparable number of task-related interactions. In both groups, group leaders are the most central actors. They supervise the work performed by their coworkers, who do not have the freedom to take major decisions on the materials to buy or on the interpretation of the fashion designers' sketches. We refer to this pattern of task communication as 'manager-centered'. Group A shows different patterns of task-related interactions among group members because of the leader's peculiar role, whose priority is to deal with

Table 3 Degree centrality in within groups task-related interactions and socially-related interactions for J, M and A

	Task-related interactions		Socially-related interactions	
Groups' members	Outdegree	Indegree	Outdegree	Indegree
J leader	352	185	40	29
J member 1	120	189	22	36
J member 2	52	99	24	18
J member 3	96	136	39	24
J (whole group)	0	13	0	18
M leader	549	315	101	25
M member 1	229	238	78	57
M member 2	290	296	108	42
M member 3	210	269	45	53
M (whole group)	0	147	0	171
A leader	72	85	17	2
A member 1	123	102	58	38
A member 2	127	158	78	54
A member 3	143	92	62	46
A (whole group)	0	30	0	75

the fashion designer: in this case, group members are given more responsibility to make decisions about the features of the collection without the leader's continuous intervention. Consequently, group A's interactions are almost homogeneously distributed among group members and the leader loses her centrality in task-related interactions. Such pattern of task communication can be labeled as 'group member-centered'.

On the contrary, an analysis of interactions classified as social shows a similar pattern for the three groups. None of the actors in the three groups seems to play a major role in the interaction flow. However, generally speaking, group J activated less social interactions (16 %) than the other two groups (respectively 21 % for M and 31 % for A). These results are consistent with Ancona's [25] external perspective: group J, which receives the highest performance rate by evaluators, is less concerned with engaging in interactions instrumental to the social life of the group, but is strongly focused on interactions instrumental to the achievement of the group goals. In addition, high levels of task interaction centralization do not seem to hamper cooperation and coordination within the group and across the group boundaries.

Interesting and counterintuitive insights derive from the analysis of the role that external communications between the three groups on one hand, and organizational actors belonging to different units on the other, play in influencing the performance evaluations that these latter give to the three groups.

In terms of the social structure that group members enacted with other organizational actors, we recorded a consistent amount of interactions taking place between the three groups and the organizational units that operate downstream the collection development process. In particular, in group A, 76 % of interactions involve outsiders, and this percentage decreases to 65 % for J and to about 58 % for M. This implies that in terms of the sole amount of external communication, the best performer group J, with approximately 1,400 interactions, seems less connected with external parties than group M (approximately 2,400 interactions) and group A (2,200 interactions). Keeping in mind that extant research supports the importance of external network ties, these results call for further elucidation.

To explain these contrasting network effects, we went back to the content of field notes to explore more deeply the role that the enacted communication network structure had on its members. Our field evidence suggests that the higher evaluations received by group J can be traced back to modalities of collaborative design not formally required and spontaneously undertaken by the group. The group does not see itself as part of a sequential process, but it promotes co-design activities, thus involving different competences from the very beginning of the collection development on. Three elements support this explanation.

In the first place, looking at the temporal distribution of interactions, it emerges that group J actively involves downstream departments since the very beginning of the project, in phases where the intervention of these departments is not formally designed by management. Figure 1 reports the distribution of the interactions activated by the three groups with downstream departments, regarding respectively communication of information/advice and orientation, request for information/advice and orientation, and signaling of a problem (we measured such distributions by splitting the whole cycle of collection development into four different phases).

J's interactions activities do not vary across the phases that compose the fashion collection. Mary, J's group leader, acknowledged during the interview how much she perceives her role as a coordinating one, even if it is not formally prescribed. In her own words:

> I realize that my work has a tremendous impact on the whole collection development process. I take decisions that will influence the whole profitability of the collection. You know, I am not sure that even top managers, like Joseph or Faust, fully grasp this aspect. We, as leaders of the collection development department, have to take decisions that are bigger than us, which have an impact that is not recognized by formal design, which doesn't favor horizontal integration. Honestly, the organizational reward system does not recognize it, either.

On the contrary, groups M and A concentrate the larger amount of interactions in the last two product development stages, when problems have already arisen and touched on downstream departments. At that point of time, the need for coordination emerges as the only way out to solve difficult situations. The following excerpt from a field note shows how the Purchasing head complains about M group leader's inability to anticipate problems and pave the way for their solution:

> The manager of the Purchasing department, Robert, starts talking with Melissa, M's group leader: 'Ok, how are you doing? Are the others aware of this situation?'
>
> Melissa: 'Well, I talked to Joseph, the COO. This morning I also briefed the Sales director'.

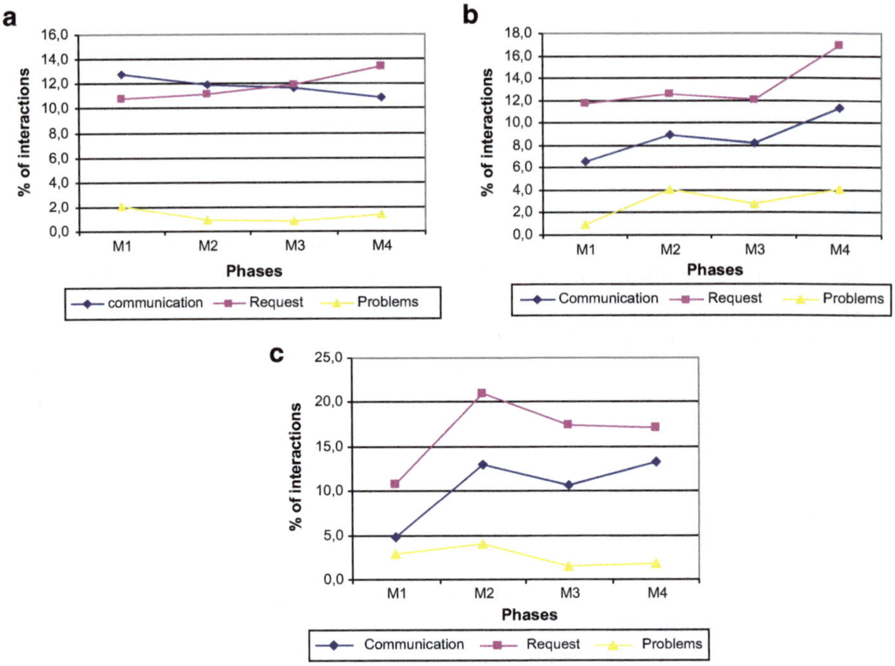

Fig. 1 (**a**) Trend of interactions over time between group J and actors external to the group for different categories of interactions. (**b**) Trend of interactions over time between group M and actors external to the group for different categories of interactions. (**c**) Trend of interactions over time between group A and actors external to the group for different categories of interactions

> Robert: 'The delay with which you ordered the fabrics is crazy. It has never happened that, at the beginning of January, we still have to launch the orders for the first collection. This is a huge problem'.
>
> Melissa: 'It is not my fault, you know that the problem is in Milan [where the designer's headquarter is located], sometimes they do not know what they want and they cause a delay'.
>
> Robert: 'Well, anyway, you should have told us about this delay; we could have managed to do something to anticipate a shortage of fabric. Problems cannot come to the surface only at the very last moment!'

Second, concerning the actual content of interactions, the analysis brings up that group J shows the lowest number of requests for information/advice/orientation, but it prompts numerous communications of information, advice, and orientation. Only few interactions have to do with signaling the presence of a problem in this group. In the majority of cases, while communicating information, J's leader communicates also the solution of a problem or elicits joint problem-solving mechanisms that enable actors to coordinate work and solve problems upon their rise. Other organizational actors do not perceive the interactions triggered by this group as an annoyance, but as a way to collaborate in the development of the best fashion collection possible. Joint problem solving allows organizational actors to work through problems while they are on the fly. As a consequence, downstream organizational actors do not have to engage in autonomous 'patching problem-solving' when issues have gone too far down the collection development process. The following excerpt from a field note tells us about a phone call from Mary to the chief of the Production department, Will, in which they jointly make decisions about flower decorations to put on an evening dress:

> Will, I have just received the flowers [that you made]. How would you like the pistils to be? Light blue, [i.e.,] same color as the fabric? Wait a minute [She takes the designer's sketches] The fabric is really, really pale blue and I think that she [the designer] meant flowers to be contrasting. Why don't we try for skin-colored, leather flowers? I think that it would be a nice fit: a striking match, but not too provocative. [She listens to Will for a minute or so] I see, I see, leather flowers, but in a matching blue. I can have a sample easily made by Alexandra [one of her group's members] by the end of today, and we can take a final step together early tomorrow morning.

Conversely, interactions touching on the other two groups A and M are perceived as a disturbance since they do not foster any cooperation (due to the time frame when they are enacted as well as to their actual content). M's members raise a large number of problems in search for a solution to outsiders, mostly in the last and more chaotic stages of the collection development process.

Eventually, J is the only group activating a significant amount of reciprocal interactions, and not just sequential ones, with other organizational units. Interactions classified as reciprocal between the groups and other organizational units amount up to 27.11 % of the total for J, 15.46 % for M and 11.95 % for A. These interactions do not imply a one-way communication, but entail the generation of a continuous sharing of information and a steady intention to jointly solve problems that might arise, thus promoting learning and innovation.

The unavailability of enacted network data would have prevented us from accounting for the role that the social networks in which groups are embedded play in shaping their effectiveness evaluations. It was mainly the actual content of the social networks that we observed that helped us explain the network effects. Our groups were similar in terms of frequency of the interactions that they played out with their environment. The difference lied in the actual content of these interactions and in their consequences on group performance. In addition to the evaluations that we specifically asked of our informants, the excerpts from the field notes below exemplify the different evaluations that organizational members provide for the three groups. In the first one, the general director congratulates group J's leader, Mary, on her group's ability to make excellent garments.

> Mary [group J's leader] goes back to look at the outfits that are on the racks and there she meets Max [general director]. They comment on each and every outfit:
>
> Max: 'Are these the army dresses that were top secret up to a couple of days ago?'
>
> Mary: 'Yes, they are.'
>
> Max: 'You rendered the gist of army clothes beautifully. I'm impressed. Everybody here would have bet that it couldn't have been done. They looked almost impossible to make. You and your group did succeed in a mission impossible-like task.'

In the second one, Joseph, the COO, complains about group A's recurrent problems with meeting deadlines and underlines that the leader does not take into account the relevance of cooperation across units:

> The COO tells Stefanie [group A's leader] with an angry voice: 'Listen, Allison [the internal designer] has told me once again that you are late with the collection delivery. That drives us all crazy. I wonder why it has always to be like that, every single collection we have to push you to work faster and to work more accurately. You aren't alone in this company: you should try much harder to solve problems with coworkers [working in different units], but you tend to stick to your office and that's it. You treat us just as firefighters.'

Eventually, the third excerpt shows how the chief of pattern makers, Florence, is dissatisfied with an output of Melissa's M group and blames her for not making earlier contacts with significant coworkers:

> Melissa: 'Unfortunately, Florence, we were obliged to put silvery brown velvet in place of light brown because we don't have that color in our storage.'
>
> Florence: 'But it makes me sick.'
>
> Melissa: 'What could I've done? That was the only fabric that we had.'
>
> Florence: 'How come, Melissa? You knew pretty well that Francine [the chief fashion designer] wanted that color badly, and you knew it ahead of time, how can you possibly not have made an order to our supplier? And then, if the right fabric was not in place, why didn't you call me when you made this jacket and ask me what my opinion about light brown was. We'd have a much better frock now if only you had given me a simple call a month ago.'

An analysis of team members' perceptions of their group's effectiveness suggests the presence of a strong alignment between inside and outside evaluations as far as the performance dimension is concerned. Group J is considered the best performer also by its internal members (mean = 5.57; S.D. 0.52), followed respectively by group A (mean = 4.86; S.D. = 0.26) and group M (mean = 4.81; S.D. 0.78). Even though the company does not run a formal performance appraisal, the constant interactions between the groups and other organizational actors can represent a vehicle for conveying appraisals and communication of satisfaction with the group throughout the company, thus favoring the above described alignment.

Conversely, concerning the group members' satisfaction with the group life, evidence shows a slightly different picture. Members of group A, that we defined as 'group member-centered', are the most satisfied (mean = 4.61, S.D. = 0.96), especially in relation to the items referred to autonomy and quality of the relationship with colleagues and supervisor. On the other hand, members of group J and group M—that we labeled as 'manager-centered'—are satisfied, but to a lesser extent (respectively J mean = 4.36, S.D. = 0.77; M mean = 4.32, S.D. = 0.69).

5 Discussion and Conclusive Remarks

As organizational achievements depend more and more on the work of groups, the comprehension of the way they function and how to improve their performance is becoming increasingly important. Our study points to the importance, while studying social networks and their effects on group performance, to consider tie content, that is to say the quality and nature of the relationship, thus going beyond the perception of the existence or lack of a particular relationship and/or its strength. Kilduff and Brass [51] have underlined lately that the research program on networks would profit from investigating more deeply the impact that different types of relations have on group functioning. Our findings suggest that it would be useful to develop a thoughtful comprehension of teams' dynamics and of their relationships within the larger organizational context. Through an ethnography conducted within three groups engaged in the development of innovative products, we show how qualitative information on the nature and dynamics of the ties between group members and other organizational actors can enhance our comprehension of the impact of network relationships on organizational behaviors.

As a consequence, we claim that the prolonged observation of group members' interactions offers researchers a privileged, thorough perspective into a group's social network. A communication network, in fact, embraces all the different ties through which organizational actors share information and advice, search and provide guidance and help, to attain organizational goals [16]. Yet, a communication network can be 'split' according to the different properties of interactions to gain multiple views and account for different courses of organizational action. As we have highlighted, a high centrality degree in the request for information/advice network as opposed to the reporting of a problem or the communication of information/advice network can generate different effects on members' behaviors and on the evaluation of groups' effectiveness. Collapsing all these dimensions of the communication network into a unique number, such as the one typically used in self-reported network data, might limit the researchers' comprehension of how network relationships really work and evolve in organizations. We believe this to be the main theoretical contribution of our study.

Our research also expands on the role of team leaders. Previous research by Balkundi et al. [9] has claimed that a formal leader's high centrality (as a proxy for prestige) in the intrateam advice network generates lower team conflict and higher team viability, whereas opposite effects are posited when the leader acts as a broker between otherwise unconnected team members. Other studies underline the positive relationship between team leaders' ties with external parties in the organizational network and team overall performance [33], team creative performance especially [52]. In particular, the latter authors suggest that leaders' gatekeeping role should be coupled with a limited involvement in the intrateam communication and problem solving network. Our findings enrich this view by stressing the positive outcomes that team leaders who have also a high external prestige, in addition to internal prestige, can bring to their groups. Contrary to Kratzer and colleagues' predictions, our data suggest the need for a 'fit' between leaders' external and internal work-related networks. Group J leader couples her external network centrality with a similar centrality in the internal network, thus suggesting that this group would be more equipped from a relational point of view to bring in and incorporate within its daily activities external inputs and insights. Conversely, group A leader acts mainly as a gatekeeper, but her peripheral role in intrateam work-related network does not allow the group to exploit the same opportunities. Group M leader's enactment of external interactions, which concentrated mainly in the second half of the collection development process, makes it hard to actively exploit external advice and information. Our evidence seems therefore to apply the argument of contingency between internal and external network ties prompted by Shan et al. [39] also to team leaders' behaviors.

In addition, our evidence brings to the fore that the experience of working with organizational actors who do not belong to the group represents an asset for the leader of the highest performing group that is not shared by other team members (who do not have equal access to external networks). An unequal spread of experience within a group adds up to the group performance: this finding is in line with Gardner et al.'s [14] argument that, when the tasks required of team members are uncertain, an uneven distribution of experiential resources increases knowledge integration capability since team members can easily detect, and make reliance on, those individuals who have more experience and competencies. Bringing these reflections together, we maintain that a team leader who is capable and willing to set and handle ties outside the team boundaries, as well as within the group, enhances the team effectiveness.

Our findings and future research should be considered in light of the study's strengths and limitations. One of its major strengths is that the research was conducted in the field on real groups working on a production task. The long time spent in the field allowed us to understand how the groups' specific tasks and networks of relationships impacted on group effectiveness. As underlined by Hansen [53] first, and by Cummings and Cross [18] later, one of the challenges in developing a deep comprehension of the relationship between groups' social structure and performance lies in the ability of researchers to come to grips with the characteristics of the work itself. Our field study has two methodological contributions to do, then. First, we integrate qualitative and quantitative techniques both in data gathering and in data analysis, while the majority of qualitative and quantitative studies are based upon respectively qualitative or quantitative techniques only. By combining qualitative, typically field note excerpts and their coding, and quantitative, mostly social network analysis techniques, we achieve a more thorough and rich comprehension of the social setting under examination. Second, unlike the vast majority of studies on social networks (e.g., [16]), observing phenomena at the time and in the place where they occur allows us to grasp the actual interactions among group members and between group members and other organizational actors, and not only those intentionally declared during structured interviews.

However, the exploratory analysis of a single case does not allow us to make general statements; accordingly, our evidence cannot be generalized to other settings. We believe that the model that we built has theoretical significance [54], but more work is needed to understand if the relations that we observed can be applied to other organizations or whether they need further refinements. Moreover, in order to assess the viability of the processes that we observed, it would be worth observing again the same groups in the development of another collection. Replication of the study in groups performing different tasks and operating in markets placing different demands on them might also help define better the interplay between social networks and group performance.

From a managerial viewpoint, the evidence of a design for manufacturing approach spontaneously activated by a group operating in a fashion industry raises important issues concerning that specific industry and has immediate practical implications for firms whose upstream activities substantially impact the ones downstream. In these cases, usually the effectiveness of first steps, which account for the product design phase, can be improved not only by developing specific design competencies, but also by linking competencies that arc distributed downstream along the production chain and, sometimes, through the involvement of suppliers and final clients in the whole process as well.

According to the above description of a fashion collection development process, it clearly appears that the tasks performed by the groups under study have the biggest impact on the whole process in terms of product quality, schedule, and budget performance. Nonetheless, only rarely do fashion firms apply design for manufacturing modalities to their manufacturing processes. The fashion industry culture in fact fosters the assumption that organizational units like engineering, production, and quality control, have to give way to the choices taken by upstream creative departments: creativity stands in the first place, no matter how this assumption may hamper the attainment of efficiency goals, and create internal tensions and conflicts [55]. The absence of constraints upon creative choices is believed to be the way to achieve high-quality products. According to this view, the collection development process seldom conveys feedbacks and recycles from downstream up to upstream departments. Under such circumstances, top management may mistakenly perceive modalities of design for manufacturing, alongside the interdependencies that it generates, as a potential threat to the very strict timetable of a fashion collection.

Nevertheless, in the context that we observed, the opportunity to map the enacted networks informed managers and group members about particular needs for communication and coordination that would have been difficult to comprehend based on self-reported networks only.

Glossary

Adjacency matrix Actor-by-actor matrix in which the cell a_{ij} reports the number of interactions activated by the actor i that involve the actor j, in relation to different properties of interactions.

Communication Network Network involving all the relationships through which organizational actors share resources such as information, help, and guide related to the execution of their work.

Group effectiveness Multidimensional construct that entails how: groups must produce outputs considered as adequate by those who receive, or who are in charge of evaluating; group members are able to work together in successive tasks; group members derive satisfaction from the execution of their tasks in the group.

Indegree The number of times an actor is drawn into an interaction by others.

Outdegree The number of times an actor initiated an interaction.

Work group Bounded set of individuals, working interdependently towards a common goal.

Workplace Social Networks Ongoing social relationships in which organizational actors are embedded in the workplace.

References

1. Zellmer-Bruhn ME (2003) Interruptive events and team knowledge acquisition. Manage Sci 49(4):514–528
2. Mathieu J, Travis Maynard M, Rapp T, Gilson L (2008) Team effectiveness 1997–2007: a review of recent advancements and a glimpse into the future. J Manage 34:410–476
3. Wageman R, Gardner H, Mortensen M (2012) The changing ecology of teams: new directions for teams research. J Organ Behav 33:301–315
4. Bell ST (2007) Deep-level composition variables as predictors of team performance: a meta-analysis. J Appl Psychol 92:595–615
5. Wageman R (1995) Interdependence and group effectiveness. Adm Sci Q 40:145–180
6. Stewart GL, Barrick MR (2000) Team structure and performance: assessing the mediating role of intrateam process and the moderating role of task type. Acad Manage J 43:135–148
7. Komaki JL, Minnich MR (2002) Crosscurrents at sea: the Ebb and flow of leadership in response to the shifting demands of racing sailboats. Group Organ Manage 27:113–140
8. Burke CS, Stagl KC, Klein C, Goodwin GF, Salas E, Halpin SM (2006) What type of leadership behaviors are functional in teams? A meta-analysis. Leadersh Q 17:288–307
9. Balkundi P, Barsness Z, Michael JH (2009) Unlocking the influence of leadership network structures on team conflict and viability. Small Group Res 40:301–322
10. Beal DJ, Cohen RR, Burke MJ, McLendon CL (2003) Cohesion and performance in groups: A metaanalytic clarification of construct relations. J Appl Psychol 88:989–1004
11. Mathieu JE, Heffner TS, Goodwin GF, Cannon-Bowers JA, Salas E (2005) Scaling the quality of teammates' mental models: equifinality and normative comparisons. J Organ Behav 26:37–56
12. Huber GP, Lewis K (2010) Cross-understanding: implications for group cognition and performance. Acad Manage Rev 35(1):6–26
13. Tesluk PE, Vance RJ, Mathieu JE (1999) Examining employee involvement in the context of participative work environments. Group Organ Manage 24:271–299
14. Gardner HK, Gino F, Staats BR (2012) Dynamically integrating knowledge in teams: transforming resources into performance. Acad Manage J 55(4):998–1022
15. Borgatti SP, Foster P (2003) The network paradigm in organizational research: a review and typology. J Manage 29:991–1013
16. Sparrowe RT, Liden RC, Wayne SJ, Kraimer ML (2001) Social network and the performance of individuals and groups. Acad Manage J 44(2):316–325
17. Reagans R, Zuckerman EW (2001) Networks, diversity, and productivity: the social capital of corporate R&D teams. Organ Sci 12(4):502–517
18. Cummings JN, Cross R (2003) Structural properties of work groups and their consequences for performance. Soc Netw 25:197–210
19. Oh H, Chung M-H, Labianca G (2004) Group social capital and group effectiveness: the role of informal socializing ties. Acad Manage J 47:860–875

20. Balkundi P, Harrison D (2006) Ties, leaders, and time in teams: strong inference about network's structure effects on team viability and performance. Acad Manage J 49(1):49–68
21. Henttonen K (2010) Exploring social network on the team level—a review of the empirical literature. J Eng Technol Manage 27:74–109
22. Brass DJ (1984) Being in the right place: a structural analysis of individual influence in an organization. Adm Sci Q 26:331–348
23. Ibarra H (1993) Network centrality, power, and innovation involvement: determinants of technical and administrative roles. Acad Manage J 36:471–501
24. Baldwin TT, Bedell MD, Johnson JL (1997) The social fabric of a team-based M. B. A program: network effects on student satisfaction and performance. Acad Manage J 40:1369–1397
25. Ancona DG (1990) Outward bound: strategies for team survival in an organization. Acad Manage J 33(2):334–365
26. Ancona DG, Caldwell DF (1992) Bridging the boundary: external activity and performance in organizational teams. Adm Sci Q 37:634–665
27. Haas MR (2002) Acting on what others know: distributed knowledge and team performance. Unpublished doctoral dissertation, Harvard University, Cambridge
28. Hackman JR (1987) The design of work teams. In: Lorsch JW (ed) Handbook of organizational behavior. Prentice-Hall, Englewood Cliffs, pp 315–342
29. Wageman R, Hackman JR, Lehman E (2005) Team diagnostic survey: development of an instrument. J Appl Behav Sci 41:373–398
30. Cross R, Parker A (2004) The hidden power of social networks. Harvard Business School Press, Boston
31. Bertolotti F, Macrì DM, Tagliaventi MR (2005) Spontaneous self-managing practices: evidence from the field. J Manage Inq 14(4):1–19
32. Wong S-S (2008) Task knowledge overlap and knowledge variety: the role of advice network structures and impact on group effectiveness. J Organ Behav 29:591–614
33. Janhonen M, Johanson J-E (2011) Role of knowledge conversion and social networks in team performance. Int J Inf Manage 31:217–225
34. Mehra A, Dixon AL, Brass DL, Robertson B (2006) The social network ties of group leaders: implication for group performance and leader reputation. Organ Sci 17(1):64–79
35. Grund TU (2012) Network structure and team performance: the case of english premier league soccer teams. Soc Netw 34(4):682–690
36. Carpenter MA, Geletkanycz MA, Sanders WG (2004) Upper echelons research revisited: antecedents, elements, and consequences of top management team composition. J Manage 30:749–78
37. Collins CJ, Clark KD (2003) Strategic human resource practices, top management team social networks, and firm performance: the role of human resource practices in creating organizational competitive advantage. Acad Manage J 46:740–751
38. Yoo JW, Reed R, Shin SJ, Lemak DJ (2009) Strategic choice and performance in late movers: influence of the top management team's external ties. J Manage Stud 46(2):308–335
39. Shan PP, Dirks KT, Chervany N (2006) The multiple pathways of high performing groups: the interaction of social networks and group processes. J Organ Behav 27:299–317
40. Chung Y, Jackson SE (2013) The external and internal networks of knowledge-intensive teams. J Manage 39(2):442–468
41. Strauss A, Corbin J (1998) Basics of qualitative research: techniques and procedures for developing grounded theory. Sage, Thousand Oaks
42. Miles MB (1983) Qualitative data as an attractive nuisance. In: Van Mannen J (ed) Qualitative methodology. Sage, Beverly Hills, pp 117–134
43. Bales RF (1950) Interaction process analysis: a method for the study of small groups. Addison-Wesley, Cambridge
44. Bales RF (1999) Social interaction systems: theory and measurement. Transaction Publishers, New Brunswick
45. Nelson RE (1988) Social network analysis as intervention tool. Group Organ Stud 13:39–58

46. Wasserman S, Faust K (1994) Social network analysis. Methods and applications. Cambridge University Press, Cambridge
47. Freeman LC (1979) Centrality in social networks: I. Conceptual clarification. Soc Netw 1:215–239
48. Borgatti SP, Everett MG, Freeman LC (2002) Ucinet for windows: software for social network analysis. Analytic Technologies, Harvard
49. Barley SR (1990) Images of imaging: notes on doing longitudinal field work. Organ Sci 3:220–247
50. Campion MA, Papper EM, Medsker GJ (1996) Relations between work team characteristics and effectiveness: a replication and extension. Personnel Psychol 49:429–452
51. Kilduff M, Brass DJ (2010) Organizational social network research: core ideas and key debates. Acad Manage Ann 4(1):317–357
52. Kratzer J, Leenders RTAJ, Van Engelen JML (2008) The social structure of leadership and creativity in engineering design teams: an empirical analysis. J Organ Behav 25:269–286
53. Hansen MT (1999) The search-transfer problem: the role of weak ties in sharing knowledge across organization subunits. Adm Sci Q 44:82–111
54. Yin RK (2003) Case study research: design and methods, 3rd edn. Sage, Thousand Oaks
55. Mills C (2011) Enterprise orientations: a framework for making sense of fashion sector start-up. Int J Entrep Behav Res 17(3):245–271

Overlapping Community Discovery Methods: A Survey

Alessia Amelio and Clara Pizzuti

Abstract The detection of overlapping communities is a challenging problem which is gaining increasing interest in recent years because of the natural attitude of individuals, observed in real-world networks, to participate in multiple groups at the same time. This review gives a description of the main proposals in the field. Besides the methods designed for static networks, some new approaches that deal with the detection of overlapping communities in networks that change over time, are described. Methods are classified with respect to the underlying principles guiding them to obtain a network division in groups sharing part of their nodes. For each of them we also report, when available, computational complexity and web site address from which it is possible to download the software implementing the method.

1 Introduction

Complex networks constitute an efficacious formalism to represent the relationships among objects composing many real world systems. Collaboration networks, the Internet, the world-wide-web, biological networks, communication and transport networks, social networks are just some examples. Networks are modeled as graphs, where nodes represent objects and edges represent interactions among these objects. One of the main problems in the study of complex networks is the detection of *community structure*, i.e. the division of a network into groups (clusters or modules) of nodes having dense intra-connections, and sparse inter-connections. In the last few years many different approaches have been proposed to uncover community structure in networks.

A. Amelio • C. Pizzuti (✉)
National Research Council of Italy (CNR), Institute for High Performance Computing and Networking (ICAR), Via Pietro Bucci, 41C, 87036 Rende (CS), Italy
e-mail: amelio@icar.cnr.it; pizzuti@icar.cnr.it

Ş. Gündüz-Öğüdücü and A.Ş. Etaner-Uyar (eds.), *Social Networks: Analysis and Case Studies*, Lecture Notes in Social Networks, DOI 10.1007/978-3-7091-1797-2_6,

Much of the effort in defining efficient and efficacious methods for community detection has been directed on finding disjoint communities. However, communities may overlap, i.e. some nodes may belong to more than one group. The membership of an entity to many groups is very common in real world networks. For example, in a social network, a person may participate in several interest groups, in collaboration networks a researcher may collaborate with many groups, in citation networks a paper could involve more than one topic, in biological networks proteins play different roles in the cell by taking part in several processes.

In this review a description of the most recent proposals for overlapping community detection is given, and a classification in different categories is provided. Algorithms have been classified by taking into account the underlying principles guiding the methods to obtain network division in groups sharing part of their nodes.

Many recent reviews describing community detection algorithms have been published [7, 11, 12, 20, 38]. However, our review differs from those of [7, 11, 12, 20] since we focus only on overlapping approaches, and from [38] because also dynamic approaches are described.

The paper is organized as follows. The next section gives some preliminary definitions necessary for the description of the approaches. Section 3 introduces the method categorization proposed in this review. Section 3.1 describes *node seeds and local expansion methods*. Section 3.2 considers *clique expansion methods*. Section 3.3 describes *link clustering algorithms*. Section 3.4 presents *label propagation approaches*. Methods for which a categorization in one of the defined classes was not possible are reported in Sect. 3.5. Section 3.6 considers the more recent proposals for dynamic networks. Section 4 gives some information regarding benchmarks that can be used to test algorithm performance. Finally, Sect. 5 gives some final considerations on the described methods and concludes the paper.

2 Preliminaries

In this section some basic definitions, necessary for a clear understanding of the concepts described in the survey, are given.

A network $\mathcal{N}$ can be modeled as a graph $G = (V, E)$ where V is a set of $n = \mid V \mid$ objects, called nodes or vertices, and E is a set of $m = \mid E \mid$ links, called edges, that connect two elements of V. A community in a network is a group of vertices (i.e. a subgraph) having a high density of edges among them, and a lower density of edges between groups. In [11] it is observed that a formal definition of community does not exist because this definition often depends on the application domain. Nevertheless, an impressive number of methods has been proposed to detect communities in complex networks. Before starting with the description of overlapping approaches, it is necessary to introduce the concept of *modularity* because of its popularity and large use among researchers.

The concept of *modularity* has been originally defined by Girvan and Newman [26] as quality function to evaluate the goodness of a partition. Along the years, however, it has been accepted as one of the most meaningful measures to partition a network, that more closely agrees with the intuitive concept of community on a wide range of real world networks.

The idea underlying modularity is that a random graph has not a clustering structure, thus the edge density of a cluster should be higher than the expected density of a subgraph whose nodes are connected at random. This expected edge density depends on a chosen *null model*. Modularity can be written in the following way:

$$Q = \frac{1}{2m} \sum_{ij} (A_{ij} - P_{ij}) \delta(C_i, C_j) \tag{1}$$

where A is the adjacency matrix of the graph G, m is the number of edges of G, and P_{ij} is the expected number of edges between nodes i and j in the null model. δ is the Kronecker function and yields one if i and j are in the same community, zero otherwise. When it is assumed that the random graph has the same degree distribution of the original graph, $P_{ij} = \frac{k_i k_j}{2m}$, where k_i and k_j are the degrees of nodes i and j respectively. Thus the modularity expression becomes:

$$Q = \frac{1}{2m} \sum_{ij} (A_{ij} - \frac{k_i k_j}{2m}) \delta(C_i, C_j) \tag{2}$$

Since only the pairs of vertices belonging to the same cluster contribute to the sum, modularity can be rewritten as

$$Q = \sum_{s=1}^{k} [\frac{l_s}{m} - (\frac{d_s}{2m})^2] \tag{3}$$

where k is the number of modules found inside a network, l_s is the total number of edges joining vertices inside the module s, and d_s is the sum of the degrees of the nodes of s. Thus the first term of each summand is the fraction of edges inside a community, and the second one is the expected value of the fraction of edges that would be in the network if edges fell at random without regard to the community structure. Values approaching 1 indicate strong community structure.

3 Methods

In this section, methods for the detection of overlapping communities are described. They have been classified in six different categories on the base of the methodology employed to identify communities. The categories are the following:

- Node seeds and local expansion
- Clique expansion

- Link clustering
- Label propagation
- Other approaches
- Dynamic networks

For each category, a short description of the main common features among algorithms of that class is provided.

3.1 Node Seeds and Local Expansion

The idea underlying these approaches is that starting from a node or a small set of nodes, a community can be obtained by adding neighboring nodes that improve a quality function. The quality function characterizes the structure of the obtained clustering.

Baumes et al. [4] introduced the concept of density function and defined a community as a subgraph that is locally optimal with respect to this density function. The internal p_{in} and external p_{ex} edge intensities of a community C are defined as

$$p_{in}(C) = \frac{E(C)}{\mid C \mid (\mid C \mid -1)} \qquad p_{ex}(C) = \frac{E_{out}(C)}{2 \mid C \mid (n- \mid C \mid)} \tag{4}$$

where $E(C)$ is the number of internal edges of C and $E_{out}(C)$ is the number of edges from nodes of C towards nodes not belonging to C. Three weight or metric functions are defined to assign a weight to a graph. The *internal edge probability*

$$W_p = p_{in}(C) \tag{5}$$

that coincides with the internal edge density, the *edge ratio*

$$W_e = \frac{E(C)}{E(C) + E_{out}(C)} \tag{6}$$

and the *intensity ratio*

$$W_i = \frac{p_{in}(C)}{p_{in}(C) + p_{ex}(C)} \tag{7}$$

These metrics measure the intensity of communication within the clusters, and can be efficiently updated when a new node is added or removed from the cluster. Finally, to measure the difference between two clusters, the Hamming, or edit distance, and the percentage of non-overlap are defined.

In order to find overlapped communities, the authors proposed two methods. The first algorithm, *Iterative Scan (IS)*, starts by choosing an edge at random, called *seed*, and adds or removes one vertex at a time until the chosen density metric improves. When no more improvement can be obtained, the algorithm stops and it is restarted with a new seed. Overlapping is possible because the restart process can reassign nodes already present in a cluster to another new forming community.

The second algorithm, *Rank Removal (RaRe)*, assumes that there are some high-ranking nodes which, when removed from the graph, disconnect the graph into smaller connected components, called cores. The deleted nodes are then added to one or more cores. This means that the overlapping between two clusters is possible only through these vertices.

To validate the methods, the Hamming distance between each cluster of the true clustering and the obtained clustering is computed, then the average of all these distances is considered.

The choice of a random edge can negatively influence the result of IS. Thus the same authors modified their *Iterative Scan* method and proposed a more efficient algorithm for finding overlapping communities [3], named IS^2, that combines *IS* and *RaRe*. *IS*^{2} also relies on a new strategy for initializing seed clusters able to compute the ranking of each node only once. The strategy is based on the observation that the only nodes capable of increasing cluster density are either the members of the cluster itself or members of the immediate neighborhood clusters, where neighboring clusters are those containing nodes adjacent to a node inside the cluster. Thus, rather than visiting each node for each iteration, nodes not belonging to one of these two groups can be skipped over.

Another variation of the IS method, that ensures connectivity of the detected communities, is described in [15].

Lancichinetti et al. [21] proposed a method to detect overlapping and hierarchical community structure, in the following referred as *LFM*, based on the concept of *fitness* of nodes belonging to a community S. Let $k_i^{in}(S)$ and $k_i^{out}(S)$ be the internal and external degrees of the nodes belonging to a community S. The *fitness* of S is then defined as

$$\mathscr{F}_S = \sum_{i \in S} \frac{k_i^{in}(S)}{(k_i^{in}(S) + k_i^{out}(S))^{\alpha}} \tag{8}$$

where α, called resolution parameter, is a positive real-valued parameter controlling the size of the communities. When $k_i^{out}(S) = 0 \ \forall i$, $\mathscr{F}_S$ reaches its maximum value for a fixed α. The community fitness has been used in [21] to find communities one at a time. *Node fitness* with respect to a community S is defined as the variation of *community fitness* of S with and without the node i, i. e.

$$\mathscr{F}_S^i = \mathscr{F}_{S \cup \{i\}} - \mathscr{F}_{S - \{i\}} \tag{9}$$

The method starts by picking a node at random, and considering it as a community S. Then a loop over all the neighbor nodes of S, not included in S, is performed in order to choose the neighbor node to be added to S. The choice is done by computing the node fitness for each node, and augmenting S with the node having the highest value of fitness. At this point the fitness of each node is recomputed, and if a node turns out to have a negative fitness value it is removed from S. The process stops when all the not yet included neighboring nodes of the nodes in S have a negative fitness. Once a community has been obtained, a new node is picked and the process restarts until all the nodes have been assigned to at least one group. The authors found that the divisions obtained for the resolution parameter $\alpha = 1$ are relevant. However, they introduce a criterion to choose a clustering based on the concept of stability. A clustering is considered stable if it is delivered for a range of values of α. The length of this range determines the more stable division, which is deemed the best result.

A different approach is applied by Lancichinetti et al. in [23] to choose the neighbor of a node to add to a cluster S. In fact, a method called *OSLOM* uses a statistical test to grow a community S by evaluating the statistical significance of a node. The intuition under the approach is that, if a vertex v shares many more edges with the nodes of S than expected in the null model, then the relationship between v and S is unexpectedly strong, thus v can be included in S.

DOCS (Detecting Overlapping Community Structures) [36] is a method based on an approach of global division followed by local expansion. *DOCS* uses classical spectral graph partitioning and random walk techniques. It first applies a spectral bi-section method with multi-level recursion to generate seed groups, and then employs a process of local expansion to add a vertex to the current cluster. The local optimization process randomly walks over the network from the seeds. In particular, the locally-optimal expansion process is based on the concept of modularity Q. At each time step, the scanned vertices are sorted in descending order with the degree normalized probabilities. If a vertex brings better Q change to the community candidates, it may be absorbed as a new member of the community structure. Furthermore, the algorithm, besides trying to achieve a good Q value, takes also into account the overlapping rate of the discovered communities. The user must fix a threshold t, and the expansion is continued until the overlapping rate is beyond the t value.

Moses [25] is a greedy algorithm that optimizes a global objective function based on a statistical network model. In this model a graph is represented by a random symmetric adjacency matrix. The objective function computes the maximum likelihood estimators from the observed likelihood. *Moses* selects an edge (u, v) at random and the community $C = \{u, v\}$ constituted by the two corresponding nodes is built by expanding it with new nodes taken from the set of neighboring nodes not yet added to C. Nodes are selected such that the objective function is maximized, and expansion continues until the highest value of the objective is obtained. Since edges are chosen at random with replacement, the same edge could be selected many times, and expanded in different communities. The algorithm periodically checks for all the communities found so far if the removal of one community improves the objective function. Furthermore, a tuning phase at the end of the expansion is performed. In this phase each node is removed from all the communities it has been assigned, and added to the communities to which it is connected by an edge. The change is accomplished only if the objective function increases. *Moses* has been compared with the methods *LFM* of Lancichinetti et al. [20], *COPRA* of Gregory [18], *Iterative Scan* (*IS*) of Baumes et al. [4], *GCE* of Lee et al. [24], *CFinder* of Palla et al. [29] . The authors performed experiments by varying the overlap to range from one to ten communities per node, i.e. a node can participate from one until ten communities. They found that when the average overlap increases, *Moses* is able to recover community structure better than the other methods.

3.2 Clique Expansion

Clique expansion methods are similar to the approaches described in the previous section. However, they consider as seed cores highly connected sets of nodes constituted by cliques, and then generate overlapping communities by merging these cliques, applying different criteria.

CFinder [1] is a system to identify and visualize overlapped, densely connected groups of nodes in undirected graphs. It also allows to navigate the original graph and the communities found. The search algorithm *CFinder* uses the *Clique Percolation Method* [29] to find *k-clique* percolation clusters. A k-clique is a complete subgraph constituted by k nodes. Two cliques are said adjacent if they share exactly $k-1$ nodes. A k-clique percolation cluster is defined as the union of all k-cliques that can be reached from each other through a series of adjacent k-cliques. The parameter k has to be provided in input. The higher the value of k, the smaller the size of the highly dense groups. The authors suggest that a value between 4 and 6 gives the richest group structure. *CFinder* works as follows. First of all, the community finding algorithm extracts all the maximal complete subgraphs, i.e. the cliques. Then a clique-clique overlap matrix is prepared. In this matrix each row (and column) corresponds to a clique, the diagonal contains the size of the clique, while the value contained in the position (i, j) is the number of common nodes between the cliques i and j. K-clique communities are obtained by deleting every element on the diagonal smaller than k, and every element off the diagonal smaller than $k-1$, replacing the remaining elements by one, and finally finding the connected components in the modified matrix.

Shen et al. [35] proposed a hierarchical agglomerative algorithm named *EAGLE* (*agglomerativE HierarchicAL clusterinG based on maximaL cliquE*) to uncover hierarchical and overlapping community structure in networks. The method is based on the concepts of *maximal cliques*, i.e. cliques that are not a subset of another clique, and *subordinate maximal cliques*, i.e. maximal cliques whose vertices are contained in some other larger maximal clique. Fixed a threshold k, all the subordinate maximal cliques with size smaller than k are discarded. The deletion of these cliques implies that some vertices do not belong to any maximal clique. These vertices are called *subordinate vertices*. The algorithm starts by considering as initial communities the maximal cliques and the subordinate vertices. Then, for each couple of communities, the similarity between them is computed and the two groups with maximum similarity are merged. This process is repeated until only one community remains. The similarity M between two communities is defined by specializing the concept of modularity, introduced by Newmann, for two communities C_1 and C_2:

$$M = \frac{1}{2m} \sum_{v \in C_1, w \in C_2, v \neq w} [A_{vw} - \frac{k_v k_w}{2m}] \tag{10}$$

where m is the total number of edges in the network, A is the adjacency matrix of the network, k_v is the degree of node v. The evaluation of the results obtained is performed by computing an extended modularity value EQ that takes into account the number of communities a node belongs to. This value is computed as follows:

$$EQ = \frac{1}{2m} \sum_i \sum_{v \in C_i, w \in C_i} \frac{1}{O_v O_w} [A_{vw} - \frac{k_v k_w}{2m}] \tag{11}$$

where O_v is the number of communities to which v belongs. The value of EQ is used by the authors to cut the generated dendrogram and to select the community structure having the maximum extended modularity value. Experiments on two real life networks show that the results obtained are meaningful.

Greedy Clique Expansion (*GCE*) is a community detection algorithm proposed by Lee at al. [24] that assigns nodes to multiple groups by expanding cliques of small size. A clique is considered as the seed or core of a community C, and it is used as starting point to obtain C by greedily adding nodes that maximize a fitness function. The fitness function adopted is that proposed by Lancichinetti et al. [20]. The algorithm first finds the maximal cliques contained in the graph G representing the network, with at least k nodes. Then it creates a candidate community C' by selecting the largest unexpanded seed and adding the nodes contained in the frontier that maximize the fitness function. This expansion is continued until the inclusion of any node would lower the fitness. At this point *GCE* checks if the community C' is near-duplicate, i.e. C' is compared with all the already obtained communities, and it is accepted only if it is sufficiently different from these communities. All the steps are repeated until no seeds remain to be considered. In order to decide if a community is near-duplicate, a distance measure between communities, based on the percentage of uncommon nodes, is introduced. This measure is defined as follows. Given two communities S and S',

$$\delta_E(S, S') = 1 - \frac{| S \cap S' |}{min(| S |, | S' |)} \tag{12}$$

This measure can be interpreted as the proportion of nodes belonging to the smaller community that are not members of larger community. Fixed a parameter ϵ as the minimum community distance, and a set of communities W, the near-duplicates of S are all the communities in W that are within a distance ϵ from S.

GCE has been compared with other state-of-the-art overlapping community detection methods, and, analogously to *Moses*, it performs better when the number of communities that a node can belong to increases from 1 to 5.

3.3 Link Clustering

Link clustering methods propose to detect overlapping communities by partitioning the set of links rather than the set of nodes. To this end, the *line graph* is used. The *line graph* $L(G)$ of an undirected graph G is another graph $L(G)$ such that (1) each vertex of $L(G)$ represents an edge of G, and (2) two vertices of $L(G)$ are adjacent if and only if their corresponding edges share a common endpoint in G. Thus a line graph represents the adjacency between edges of G. The main advantage of applying clustering to the line graph is that it produces an overlapping graph division of the original interaction graph, thus allowing nodes to be present in multiple communities.

Pereira et al. [30] have been the first in using the line graph to find overlapping modules for protein–protein interaction networks. To this end they applied a well known method (MCL [8]) for protein interaction networks to the line graph.

Evans and Lambiotte [10] argued that any algorithm that partitions a network can be applied to partition links for discovering overlapping community structure. They first reviewed a definition of modularity which uses statistical properties of dynamical processes taking place on the edges of a graph, and then proposed link partitioning by applying the new modularity concept. In particular, the traditional modularity Q is defined in terms of a random walker moving on the links of the network. Such a walker would therefore be located on the links instead of the nodes at each time t, and its movements are between adjacent edges, i.e. links having one node in common. Three quality functions for partitioning links of a network G have been proposed. Each formalizes a different dynamical process and explores the structure of the original graph G in a different way.

In the first dynamical process, *Link-Link random walk*, the walker jumps to any of the adjacent edges with equal probability. In the second process, *Link-Node-Link random walk*, the walker moves first to a neighboring node with equal probability, and then jumps to a new link, chosen with equal probability from those new edges incident at the node. In the last process, the dynamics are driven by the original random walk but are projected on the links of the network. The stabilities of the three processes have been defined by generalizing the concept of modularity to paths of arbitrary length, in order to tune the resolution of the optimal partitions. The optimal partitions of these quality functions can be discovered by applying standard modularity optimization algorithms to the corresponding line graphs. An extension to deal with weighted line graphs is reported in [9].

GA-NET+ is a method proposed by Pizzuti [31] that employs *Genetic Algorithms* [14]. A Genetic Algorithm (*GA*) evolves a constant-size population of elements (called *chromosomes*) by using the genetic operators of *reproduction*, *crossover* and *mutation*. Each chromosome represents a candidate solution to a given problem and it is associated with a *fitness value* that reflects how good it is, with respect to the other solutions in the population. The method uses the concept of *community score* to measure the quality of the division in communities of a network. Community score is defined as follows.

Let μ_i denote the fraction of edges connecting node i to the other nodes in a community S. More formally

$$\mu_i = \frac{1}{\mid S \mid} k_i^{in}(S)$$

where $\mid S \mid$ is the cardinality of S, and $k_i^{in}(S) = \sum_{j \in S} A_{ij}$ is the number of edges connecting i to the other nodes in S.

The *power mean of S of order r*, denoted as $\mathbf{M}(S)$ is defined as

$$\mathbf{M}(S) = \frac{\sum_{i \in S}(\mu_i)^r}{|S|} \tag{13}$$

The *volume* v_S of a community S is defined as the number of edges connecting vertices inside S, i.e the number of 1 entries in the adjacency sub-matrix of A corresponding to S, $v_S = \sum_{i,j \in S} A_{ij}$.

The *score* of S is defined as $score(S) = \mathbf{M}(S) \times v_S$. The *community score* of a clustering $\{S_1, \ldots S_k\}$ of a network is defined as

$$\mathscr{C}\mathscr{S} = \sum_{i}^{k} score(S_i) \tag{14}$$

Community score gives a global measure of the network division in communities by summing up the local score of each module found. The problem of community identification can then be formulated as the problem of maximizing $\mathscr{C}\mathscr{S}$.

The algorithm tries to maximize $\mathscr{C}\mathscr{S}$ by running the genetic algorithm on the line graph $L(G)$ of the graph G modeling the network. As already pointed out, a main advantage in using the line graph is that the partitioning of $L(G)$ obtained by *GA-NET+* corresponds to an overlapping graph division of G. The method uses the locus-based adjacency representation. In this representation an individual of the population consists of N genes $g_1, \ldots, g_N$ and each gene can assume allele values j in the range $\{1, \ldots, N\}$. Genes and alleles represent nodes of the graph $G = (V, E)$ modelling a network $\mathscr{N}$, and a value j assigned to the ith gene is interpreted as a link between the nodes i and j of V, thus in the clustering solution found i and j will be in the same cluster. *GA-NET+* starts by generating a population initialized at random with individuals representing a partition in subgraphs of the line graph $L(G)$. After that, the fitness of the individuals from the original graph must be evaluated and a new population of individuals is created by applying uniform crossover and mutation. The dense communities present in the network structure are obtained at the end of the algorithm, without the need to know in advance the exact number of groups. This number is automatically determined by the optimal value of the community score.

Ahn et al. [2] proposed a hierarchical agglomerative link clustering method to group links into topologically related clusters. The algorithm applies a hierarchical method to the line graph by defining two concepts: *link similarity* and *partition density*. Link similarity is used during the single-linkage hierarchical method to find the pair of links with the largest similarity in order to merge their respective communities. This similarity measure is defined as follows. Let a node i be given, the *inclusive neighbors* of i are

$$n_+(i) = \{x \mid d(i,x) \leq 1\} \tag{15}$$

where $d(i,x)$ is the length of the shortest path between nodes i and x. Thus $n_+(i)$ contains the node itself and its neighbors. Then, the similarity S between two links e_{ik} and e_{jk} can be defined by using the Jaccard index:

$$S(e_{ik}, e_{jk}) = \frac{\mid n_+(i) \cap n_+(j) \mid}{\mid n_+(i) \cup n_+(j) \mid} \tag{16}$$

The similarity between links is also extended to networks with weighted, directed, or signed links (without self-loops). The agglomerative process is repeated until all links belong to a single cluster. To find a meaningful community structure, it is necessary to decide where the built dendrogram must be cut. To this end, the authors introduce a new quantity, the *partition density D*, that measures the quality of a link partitioning.

Partition density is defined as follows. Let m be the number of links of a given network, and $\{P_1, \ldots, P_C\}$ the partition of the links in C subsets. Each subset P_c has $m_c =\mid P_c \mid$ links and $n_c =\mid \cup_{e_{ij} \in P_c} \{i, j\} \mid$ nodes. Then

$$D_c = \frac{m_c - (n_c - 1)}{n_c(n_c - 1)/2 - (n_c - 1)} = 2\frac{m_c - (n_c - 1)}{(n_c - 2)(n_c - 1)} \tag{17}$$

is the normalization of the number of links m_c by the minimum and maximum number of possible links between n_c connected nodes. It is assumed that $D_c = 0$ when $n_c = 2$. The partition density D is the average of the D_c, weighted by the fraction of present links:

$$D = \frac{2}{m} \sum_c m_c \frac{m_c - (n_c - 1)}{(n_c - 2)(n_c - 1)} \tag{18}$$

The authors, in order to compare their approach with other state-of-the-art methods, introduce four measures. Two measures, *community quality* and *overlap quality*, are based on metadata known possessed by some networks studied in the literature. These metadata consist of a small set of annotations or tags attached to each node.

The other two measures, *community coverage* and *overlap coverage*, consider the amount of information extracted from the network. Community coverage counts the fraction of nodes that belong to at least one community of three or more nodes, called nontrivial communities. The authors state that this measure provides a sense of how much of the network is analyzed. Overlap coverage counts the average number of membership per nodes to nontrivial communities. When the communities are not overlapping, the two coverage measures give the same information.

3.4 Label Propagation

In label propagation approaches a community is considered a set of nodes which are grouped together by the propagation of the same property, action or information in the network.

Gregory in [19] proposed the algorithm *COPRA* (*Community Overlap PRopagation Algorithm*), as an extension of the label propagation technique of Raghavan et al. [32]. The main modification consists in assigning multiple community identifiers to each vertex. The method, thus, associates with each vertex x a set of couples (c, b), where c is a community identifier and b is a belonging coefficient expressing the strength of x as member of community c. *COPRA* starts by giving to each vertex a single label with belonging coefficient set to 1. Then, repeatedly, each vertex x updates its labels by summing and normalizing the belonging coefficients of its neighboring nodes. The new set of x's labels is constituted by the union of its neighbor labels. However, in order to limit the number of communities a vertex can participate, a parameter v must be given in input. In particular, the labels whose belonging coefficient is less that $1/v$ are deleted. *COPRA* has a nondeterministic behavior when all the belonging coefficients corresponding to the labels associated with a vertex are the same, but below the threshold. In such a case a randomly selected label is maintained, while the remaining are discarded. Finally, communities totally contained in others are removed, and disconnected communities that could be generated are split in connected ones.

Wu et al. in [37] pointed out that the input parameter v makes *COPRA* unstable since it is a global vertex-independent parameter not taking into account that, often, most nodes are non-overlapping, while few nodes participate in many communities. Thus an appropriate choice of v is difficult and it induces the non-determinism described above. To overcome this shortcoming, Wu et al. proposed *BMLPA* (*Balanced Multi-Label Propagation Algorithm*), a method based on a new label update strategy that computes balanced belonging coefficients, and does not limit the number of communities a node can belong to. A balanced belonging coefficient is computed by normalizing each coefficient by the maximum value a vertex has, and retaining it only if its normalized value is above a fixed threshold p. Another characteristic introduced by *BMLPA* is the initialization process of vertex labels based on the extraction of overlapping rough cores. Such cores allow *BMLPA* to efficiently assign labels to each node, and to effectively update labels since the threshold p, independently the value chosen, would make the new update strategy not work well because each node would retain all the labels of its neighbors.

SLPA [39, 40] is another extension of the label propagation technique of Raghavan et al. [32] that adopts a speaker–listener based information propagation process. Each node is endowed with a memory to store the labels received. It can have both the role of *listener* and *speaker*. In the former case it takes labels from the neighbors and accepts only one following a listening rule, such as the most popular observed at the current step. If it is a speaker, it sends a label to the neighboring listener node by choosing a label with respect to a certain speaker rule, such as single out a label with probability proportional to its frequency in the memory. The algorithm stops when a fixed number t of iterations has been reached.

3.5 Other Approaches

In this section methods which could not be categorized in one of the above classes are reported.

Zhang et al. [41] developed an algorithm for detecting overlapping community structure by combining modularity concept, spectral relaxation and fuzzy c-means. In particular, a new modularity function extending the Newman's modularity concept is introduced to take into account soft assignments of nodes to communities. The problem of maximizing the modularity function is reformulated as an eigenvector problem. Fixed an upper bound to the number k of communities, the top $k-1$ eigenvectors of a generalized eigen system are computed, and a mapping of the network nodes into a d-dimensional Euclidean space is performed, where $d \leq k-1$. After that, fuzzy c-means clustering is applied to group nodes by maximizing the modified modularity function.

In [16] Gregory presented a hierarchical, divisive approach, based on Girvan and Newman's algorithm (GN) [13], but extended with a novel method of splitting vertices. *CONGA* (*Cluster-Overlap Newman Girvan Algorithm*) adds to the GN algorithm the possibility to split vertices between communities, based on the concept of *split betweenness*. This concept allows to choose either to split a vertex or remove an edge. The edge betweenness of an edge e is the number of shortest paths, between all pairs of vertices, that pass along e. A high betweenness indicates that the edge acts as a bottleneck between a large number of vertex pairs and suggests that it is an inter-cluster edge. The split betweenness of a vertex v is the number of shortest paths that would pass between the two parts of v if it were split. A vertex can be split in many ways; the best split is the one that maximizes the split betweenness. An approximate, efficient algorithm has been presented for computing split betweenness and edge betweenness at the same time. In *CONGA*, a network is initially considered as a single community, assuming it is connected. After one or more iterations, the network is subdivided into two components (communities). Communities are repeatedly split into two until only singleton communities remain.

If binary splits are represented as a dendrogram, the network can be divided into any desired number of communities.

CONGA has a complexity of $O(m^3)$, where m is the number of edges, thus it is rather inefficient. A faster implementation of *CONGA*, named *CONGO*, that uses local betweenness and runs in $O(m \log m)$ is proposed by the same author in [17].

Chen et al. in [6] proposed an approach to find communities with overlaps and outlier nodes based on visual data mining. They consider a community as a network partition whose entities share some common features and a relationship metric is adopted to evaluate the proximity of the entities each other. Such metric is based on the notion of random connections useful to identify communities which are considered as non-random structures. Furthermore, it takes into account the neighborhood around any two nodes in order to evaluate their relationship. An algorithm that generates an ordering of the network nodes according to their relation scores is presented. From this ordered list of nodes, communities are obtained by considering a consecutive group of nodes with high relation score. A 2D visualization of these scores shows peaks and valleys, where a sharp drop of relation scores after a peak is interpreted as the end of a community, while the valleys between two peaks represent a set of hubs which belong to several communities. By this visualization a user is requested to fix a *community threshold* and an *outlier threshold* that allow the algorithm to decide whether a node should be considered an outlier or be added to the current community. The authors state that the main advantage of this visual mining approach is that a user can easily provide input parameters that allow the method to find communities, hub nodes, and outliers.

Rees and Gallagher [33] proposed an approach to discover communities based on the collective viewpoint of individuals. The base concept is that each node in the network knows, by way of its *egonet*, the members of its friendship group. An egonet is an induced subgraph composed of a central node, its neighbors, and all edges among nodes in the egonet that are also links in the main graph. Therefore, by merging each individual's views of friendship groups, communities can be discovered. The friendship groups represent the small clusters, extracted from egonets, composed of the central node and connected neighbors. More friendship-groups can be combined to create a community. The algorithm consists of two steps; the first one is the detection of friendship groups, while the second step comprises the merging of friendship groups into communities. In step one, the algorithm iterates through every node in the graph, centering on the selected vertex and computing the egonet. After that, friendship groups are extracted from that egonet, and the central node is eliminated, since it is known to exist in multiple friendship groups. Consequently, the graph breaks into multiple connected components. The central vertex is then added back to each connected component obtained to create the friendship groups. The output of the first phase is a set of friendship groups, from an egocentric point of view. The second step consists in merging the groups into communities. This process is done by first merging all exact matches, i.e. groups that are complete or proper subsets of other groups. Finally, groups that are relatively close, i.e. groups that match all but one element from the smaller group, are merged. The process is repeated until no more merges can be performed.

The same authors, in [34] presented a *swarm intelligence* approach for overlapping community detection. A network is considered as a set of agents corresponding to nodes characterized by neighbors. Each agent interacts with its social groups. The agent knows the set of its friends and even which of their friends are also friends. Friends agree on a common community ID inside the different social groups or friendship groups. The algorithm consists of the following steps. First of all, each agent is assigned an identifier ID. It determines a complete map of its neighborhood and builds an egonet. Starting from the egonet, the friendship groups can be extracted by a union-find algorithm. Each friendship group is identified by a unique ID, composed by the base agent ID and a unique incrementing decimal value. Secondly, within each friendship group, the agent will ask to its neighbors their views of the friendship group (which can be various from different perspectives) in order to identify the non-propagating nodes (nodes whose views differ from the agent view and consequently their information is not further propagated). Finally, the assigned friendship groups IDs are propagated. In particular, if the ID value on one of the friendship groups has been modified, the new ID value will be spread to the nodes inside the friendship group. The process is repeated until the convergence in propagation has been reached. At the end of the process, each agent will have a list of assigned communities. Communities can be easily detected because they are groups of agents that share a common ID value.

3.6 Dynamic Networks

The methods described so far do not take into account an important aspect characterizing networks: i.e. the evolution they go through over time. The representation of many complex systems through a static graph, even when the temporal dimension describing the varying interconnections among nodes is available, does not allow to study the network dynamics and the changes it incurs over time.

Dynamic networks, instead, capture the modifications of interconnections over time, allowing to trace the changes of network structure at different time steps. Analyzing networks and their evolution is recently receiving an increasing interest from researchers. However, there have been few proposals for the detection of overlapping communities in dynamic social networks. In the following the more recent methods aiming to seek out dynamic communities are described.

Palla et al. [28] have been among the first researchers to introduce an approach that allows to analyze the time dependence of overlapping communities on a large scale and as such, to uncover basic relationships characterizing community evolution. Actually they argued that at each time step communities can be extracted by using the *Clique Percolation Method (CPM)* [29]. The events that characterize the life time of a community are growth or contraction, at each time step a new community can appear, while others can disappear. Furthermore groups can merge or split. In order to identify community evolution along time, the authors proposed to merge networks of two consecutive time steps t and $t+1$, and then apply the *CPM* method to extract the new community structure of the joint network. Since the joined graph contains the union of the links of the two graphs, any community from time step t to $t+1$ can only grow, merge or remain unchanged, more than one community can be merged into a single community, but no community may loose members. Consequently, any community in one of the original networks can be contained in exactly one community of the joined network. Communities in the joint graph provide a way to match communities between time steps t and $t+1$. If a community in the joint graph contains a single community from t and a single community from $t+1$, then they are matched. If the joint group contains more than one community from either time steps, the communities are matched in descending order of their relative node overlap. Overlap is computed for every pair of communities from the two time steps as the fraction of the number of common nodes to the sum of number of nodes of both communities. Experiments on two real-life networks showed that large groups remain alive if they undergo dynamic changes. On the contrary, small groups survive if they are stable.

Cazabet et al. [5] introduced the concepts of *intrinsic community* and *longitudinal detection*, and proposed an algorithm named *iLCD* (*intrinsic Longitudinal Community Detection*) to discover highly overlapping groups of nodes. An intrinsic community is considered one that owns a characteristic deemed meaningful, such as for example being a 4-clique. Longitudinal detection means that, starting from an intrinsic community, new members join gradually like the snowball effect. *iLCD* considers the list of edges ordered with respect to the time they appeared, and, for each edge(u, v) of the set of edges E_t created at time t, it performs three steps. First, for each community C which u (resp. v) belongs to, it tries to add v (resp. u), as explained below. Then, if u and v do not already belong to any community, it tries to create a new one; finally, similar communities are merged. The first step of updating existing communities by the addition of new nodes is realized by estimating, for each community, the mean number of second neighbors *EMSN*, i.e. nodes that can be reached with a path of length 2 or less, and the mean number of robust second neighbors *EMRSN*, i.e. nodes that can be reached with at least two paths of length 2 or less. A new node is accepted in the community if the number of its neighbors at rank 2 is greater than *EMSN*, or the number of its robust neighbors at rank 2 is greater than *EMRSN*. The second step of creating a new community checks whether the couple of nodes (u, v) constitutes a minimal predefined pattern, like a 4-clique. This intrinsic property must be predetermined. Finally merging is executed by fixing an overlap threshold, and when two communities have an overlap above the threshold, the smaller is deleted and the greater is retained. This choice, as the authors point out, limits the use of uncertain heuristics. Comparison with other methods shows that this approach outperforms *CPM* only if the network is highly dense.

Another recent proposal to detect overlapping communities in dynamic networks is described in [27] by Nguyen et al. The method, named *AFOCS* (*Adaptive Finding Overlapping Community Structure*), consists of two phases. In the first phase local communities are obtained by searching for all the groups of nodes C whose internal density

$$\Psi(C) = \frac{| C^{in} |}{| C | * (| C | - 1)/2} \tag{19}$$

where C^{in} is the number of internal connections of C, i.e. the number of links having both endpoints in C, is higher than a threshold $\tau(C)$ defined as

$$\tau(C) = \frac{\sigma(C)}{| C | * (| C | - 1)/2} \tag{20}$$

where

$$\sigma(C) = (| C | * (| C | - 1)/2)^{1 - \frac{1}{|C| * (|C| - 1)/2}} \tag{21}$$

The local communities are then merged provided that their overlapping score is higher than a value given as input parameter. The second phase adaptively updates the communities obtained in the first step, by considering how the network evolves over time. The authors individuate four major changes a network can incur: a new node and its adjacent edges are either added or removed to/from the network; a new edge connecting two existing nodes is added or an existing edge is removed. The algorithm is able to obtain the new community structure by adopting the more apt strategy to determine whether a community will split, or two communities will merge. A comparison with existing approaches showed that *AFOCS* performances are competitive with other methods, mainly as regards running time.

4 Benchmarks for Testing Algorithms

The capability of an algorithm in detecting community structure is usually validated by testing the method on artificial or real world networks for which the division in communities is known. Since the availability of ground-truth community structure for large real networks is rather difficult, synthetic benchmarks built by specifying parameters to characterize network structure are preferred.

One of the most known benchmarks for non-overlapping networks has been proposed by Girvan and Newan in [13]. The network consists of 128 nodes divided into four communities of 32 nodes each. Edges are placed between vertex pairs at random but such that $z_{in} + z_{out} = 16$, where z_{in} and z_{out} are the internal and external degree of a node with respect to its community. If $z_{in} > z_{out}$ the neighbors of a node inside its group are more than the neighbors belonging to the other three groups, thus a good algorithm should discover them. This benchmark, as observed in [20], however, is rather simple since it is characterized by non-overlapping communities having all the same size and by nodes having the same expected degree. Thus, Lancichinetti et al. [22] proposed a new class of benchmarks that extends the Girvan and Newman's benchmark by introducing power law degree distributions, different community size, and percentage of overlap between communities.

Table 1 A summarization of the reviewed methods. For EAGLE, h is the number of pairs of maximal cliques which are neighbors, and s is the number of maximal cliques; for GCE, h is the number of cliques; for Ahn and AFOCS, d_{max} is the maximum node degree; for GA-NET+ t is the number of generations and p the population size; for SLPA t is the number of iterations performed by the algorithm

Approach	Method	Reference	Complexity
Node seeds and local expansion	IS, RaRe	Baumes et al.	
	IS^2, *CIS*	[3,4,15]	$O(mk+n)$
	LFM	Lancichinetti et al.	
	OSLOM	[21,23]	$O(n^2)$
	DOCS	Wei et al. [36]	–
	Moses	McDaid and Hurley [25]	$O(n^2)$
Clique expansion	CFinder	Palla et al.	$O(m^{\frac{ln\ m}{10}})$
	CPM	[1,29]	
	EAGLE	Shen et al. [35]	$O(n^2+(n+h)s)$
	GCE	Lee et al. [24]	$O(mh)$
Line graph	Evans	Evans and Lambiotte [9,10]	$O(2mk\ log\ n)$
	GA-NET+	Pizzuti [31]	$O(tp(m+m\ log\ m)$
	Ahn	Ahn et al. [2]	$O(n\,d_{max}^2)$
Label propagation	COPRA	Gregory [19]	$O(vm\ log(vm/n))$
	BMLPA	Wu et al. [37]	$O(n\ log\ n)$
	SLPA	Xie et al. [39,40]	$O(tm)$
Dynamic methods	CPM	Palla et al. [29]	
	iLCD	Cazabet et al. [5]	$O(nk^2)$
	AFOCS	Nguyen et al. [27]	$O(d_{max}m)+O(n^2)$
Other methods	FCM	Zhang et al. [41]	$O(nk^2)$
	CONGA	Gregory [16]	$O(m^3)$
	CONGO	Gregory [17]	$O(m\ log\ m)$
	ONDOCS	Chen et al. in [6]	–
	Egonet	Rees and Gallagher [33]	$O(n(log\ n)^2+n^2\ log\ n)$
	Swarm egonet	Rees and Gallagher [34]	$O(n\ log^2\ n)$

The benchmark is also characterized by the *mixing parameter* $\mu = \frac{z_{out}}{z_{in}+z_{out}}$ that gives the ratio between the external degree of a node and the total degree of the node. When $\mu < 0.5$ the communities are well defined, thus a good algorithm should discover them. The software is public and can be downloaded from http://sites.google.com/site/andrealancichinetti/software.

5 Discussion and Conclusions

The paper reviewed state-of-the-art approaches for the detection of overlapped communities. Methods have been classified in different categories. Node seed and local expansion methods together with clique expansion approaches are characterized by

Table 2 Methods for which it is possible to download the software

Reference	Software
Lancichinetti et al. [21, 23]	https://sites.google.com/site/andrealancichinetti/software
McDaid and Hurley [25]	https://sites.google.com/site/aaronmcdaid/downloads
Palla et al. [1, 29]	http://www.cfinder.org
Lee et al. [24]	https://sites.google.com/site/greedycliqueexpansion
Evans and Lambiotte [9, 10]	https://sites.google.com/site/linegraphs/
Pizzuti [31]	http://www.icar.cnr.it/pizzuti/codes.html
Ahn et al. [2]	https://github.com/bagrow/linkcomm
Gregory [19]	http://www.cs.bris.ac.uk/~steve/networks/copra/
Wu et al. [37]	http://dev.bjtu.edu.cn/bmlpa/
Xie et al. [39, 40]	https://sites.google.com/site/communitydetectionslpa
Cazabet et al. [5]	http://cazabetremy.fr/Cazabet_remy/iLCD.html
Gregory [16]	http://www.cs.bris.ac.uk/~steve/networks/congapaper/
Gregory [17]	http://www.cs.bris.ac.uk/~steve/networks/congopaper/

the same idea of starting with a node (in the former case), or a group of nodes (in the latter case) and then expanding the current cluster until the adopted quality function increases. Link clustering methods detect overlapping communities by partitioning the set of links rather than the set of nodes by using the *line graph* corresponding to the network under consideration. Since generally the number of edges is much higher than the number of nodes, these methods are computationally expensive. Label propagation approaches are among the most efficient since they start from a node and visit neighboring nodes to propagate class label. Approaches reported in Sect. 3.5 to find overlapping communities adopt strategies that substantially differ from the others. Thus Zhang et al. [41] use modularity, spectral relaxation and fuzzy c-means, Gregory [17] relies on the concept of split betweenness to duplicate a node, Chen et al. [6] propose an interactive approach based on visual data mining, Rees and Gallagher [33, 34] use the concept of egonet and apply swarm intelligence. Dynamic approaches try to deal with the problem of network evolution and constitute a valid help in understanding changes a network might undergo over time.

A summarization of the described methods is reported in Table 1. For each method, when known, the computational complexity is reported. Furthermore, in Table 2, a link to the web site from which it is possible to download the software implementing the algorithm is given.

Though the number of approaches present in the literature is notably, the results obtained by each of them are substantially different, thus there is no a universal method that is competitive with respect to all the others for networks having different characteristics such as sparsity, degree distribution, overlap percentage among communities, and so on. As pointed out in [38], there are two questions that researchers should focus on: "when to apply overlapping methods and how significant the overlapping is". Investigation on these issues and extensions to weighted networks constitute open problems for future research.

References

1. Adamcsek B, Palla G, Farkas IJ, Derényi I, Vicsek T (2006) Cfinder: locating cliques and overlapping modules in biological networks. Bioinformatics 22:1021–1023
2. Y-Y Ahn, Bagrow JP, Lehmann S (2010) Link communities reveal multiscale complexity in networks. Nature 466:761–764
3. Baumes J, Goldberg M, Magdon-Ismail M (2005) Efficient identification of overlapping communities. In: Proceedings of the 2005 IEEE international conference on intelligence and security informatics (ISI'05). Springer, Berlin, pp 27–36
4. Baumes J, Goldberg MK, Krishnamoorthy MS, Magdon-Ismail M, Preston N (2005) Finding communities by clustering a graph into overlapping subgraphs. In: IADIS AC. IADIS, pp 97–104
5. Cazabet R, Amblard F, Hanachi C (2010) Detection of overlapping communities in dynamic social networks. In: IEEE international conference on social computing/IEEE conference on privacy, security, risk and trust, pp 309–314
6. Chen J, Zaiane OR, Sander J, Goebel R (2010) Ondocs: ordering nodes to detect overlapping community structure. In: Memon N, Xu JJ, Hicks DL, Chen H (eds) Data mining for social network data. Annals of information systems, vol 12. Springer, New York, pp 125–148
7. Coscia M, Giannotti F, Pedreschi D (2011) A classification for community discovery methods in complex networks. Stat Anal Data Min 5(4):512–546
8. Enright AJ, Dongen SV, Ouzounis CA (2002) An efficient algorithm for large-scale detection of protein families. Nucleic Acids Res 30(7):1575–84
9. Evans TS, Lambiotte R (2010) Line graphs of weighted networks for overlapping communities. Eur Phys J B 77(2):265–272
10. Evans TS, Lambiotte R (2009) Line graphs, link partitions, and overlapping communities. Phys Rev E 80(1):016105:1–016105:8
11. Fortunato S (2010) Community detection in graphs. Phys Rep 486:75–174
12. Fortunato S, Castellano C (2007) Community structure in graphs (arXiv:0712.2716v1 [physics.soc-ph])
13. Girvan M, Newman MEJ (2002) Community structure in social and biological networks. Proc Natl Acad Sci USA 99:7821–7826
14. Goldberg DE (1989) Genetic algorithms in search, optimization, and machine learning. Addison-Wesley, Reading
15. Goldberg M, Kelley S, Magdon-Ismail M, Mertsalov K, Wallace A (2010) Finding overlapping communities in social networks. In: Proceedings of the 2010 IEEE second international conference on social computing (SOCIALCOM '10), pp 104–113
16. Gregory S (2007) An algorithm to find overlapping community structure in networks. In: Proceedings of the 11th European conference on principles and practice of knowledge discovery in databases, PKDD 2007. Springer, Berlin, pp 91–102
17. Gregory S (2008) A fast algorithm to find overlapping communities in networks. In: Proceedings of the 12th European conference on principles and practice of knowledge discovery in databases, PKDD 2008. Springer, Berlin, pp 408–423
18. Gregory S (2009) Finding overlapping communities using disjoint community detection algorithms. In: Fortunato S, Mangioni G, Menezes R, Nicosia V (eds) Complex networks. Studies in computational intelligence, vol 207. Springer, Berlin, pp 47–61
19. Gregory S (2010) Finding overlapping communities in networks by label propagation. New J Phys 12(10):103018
20. Lancichinetti A, Fortunato S (2009) Community detection algorithms: a comparative analysis. Phys Rev E 80:056117
21. Lancichinetti A, Fortunato S, Kertész J (2009) Detecting the overlapping and hierarchical community structure of complex networks. New J Phys 11:033015
22. Lancichinetti A, Fortunato S, Radicchi F (2008) Benchmark graphs for testing community detection algorithms. Phys Rev E 78:046110

23. Lancichinetti A, Radicchi F, Ramasco JJ, Fortunato S (2011) Finding statistically significant communities in networks. PLOS one 6:e18961
24. Lee C, Reid F, McDaid A, Hurley N (2010) Detecting highly overlapping community structure by greedy clique expansion. In: Workshop on social network mining and analysis
25. McDaid A, Hurley N (2010) Detecting highly overlapping communities with model-based overlapping seed expansion. In: Proceedings of the 2010 international conference on advances in social networks analysis and mining (ASONAM '10), pp 112–119
26. Newman MEJ, Girvan M (2004) Finding and evaluating community structure in networks. Phys Rev E69:0261135
27. Nguyen NP, Dinh TN, Tokala S, Thai MT (2011) Overlapping communities in dynamic networks: their detection and mobile applications. In: Proceedings of the 17th annual international conference on mobile computing and networking (MOBICOM 2011), pp 85–96
28. Palla G, Barabasi A, Vicsek T (2007) Quantifying social group evolution. Nature 446:664–667
29. Palla G, Farkas IJ, Derényi I, Vicsek T (2005) Uncovering the overlapping community structure of complex networks in nature and society. Nature 435:814–818
30. Pereira JB, Enright AJ, Ouzounis CA (2004) Detection of functional modules from protein interaction networks. Proteins Struct Funct Bioinforma 54(1):49–57
31. Pizzuti C (2009) Overlapped community detection in complex networks. In: Proceedings of the 11th annual conference on genetic and evolutionary computation (GECCO '09), pp 859–866
32. Raghavan UN, Albert R, Kumara S (2007) Near linear time algorithm to detect community structures in large-scale networks. Phys Rev E 76:036106
33. Rees BS, Gallagher KB (2010) Overlapping community detection by collective friendship group inference. In: Proceedings of the 2010 international conference on advances in social networks analysis and mining (ASONAM '10), pp 375–379
34. Rees BS, Gallagher KB (2012) Overlapping community detection using a community optimized graph swarm. Soc Netw Anal Min 2(4):405–417
35. Shen H, Cheng X, Cai K, Hu M-B (2009) Detect overlapping and hierarchical community structure in networks. Phys A Stat Mech Appl 388(8):1706–1712
36. Wei F, Qian W, Wang C, Zhou A (2009) Detecting overlapping community structures in networks. World Wide Web 12(2):235–261
37. Wu Z-H, Lin Y-F, Gregory S, Wan H-Yu, Tian S-F (2012) Balanced multi-label propagation for overlapping community detection in social networks. J Comput Sci Technol 27(3):468–479
38. Xie J, Kelley S, Szymanski BK (2013) Overlapping community detection in networks: the state of the art and comparative study. ACM Comput Surv 45(4) (Art No. 43)
39. Xie J, Szymanski BK (2012) Towards linear time overlapping community detection in social networks. In: Advances in knowledge discovery and data mining - 16th Pacific-Asia conference (PAKDD 2012), pp 25–36
40. Xie J, Szymanski BK, Liu X (2011) Slpa: uncovering overlapping communities in social networks via a speaker-listener interaction dynamic process. In: Proceedings of ICDM workshops on data mining technologies for computational collective intelligence, pp 344–349
41. Zhang S, Wang R-S, Zhang X-S (2007) Identification of overlapping community structure in complex networks using fuzzy -means clustering. Phys A Stat Mech Appl 374(1):483–490

Classification in Social Networks

Zehra Çataltepe and Abdullah Sönmez

Abstract Production of social network data in different kinds and huge amounts brings with it classification problems which need to be solved. In this chapter, we introduce a framework for classification in a social network. Aggregation of neighbor labels and sampling are two important aspects of classification in a social network. We give details of different aggregation methods and sampling methods. Then, we discuss different graph properties, especially homophily, which may be helpful in determining which type of classification algorithm should be used. We give details of a collective classification algorithm, ICA (Iterative Classification Algorithm), which can be used for semi-supervised learning in general and transductive learning in particular on a social network. We present classification results on three different datasets, using different aggregation and sampling methods and classifiers.

1 Introduction

In most machine learning applications, the observed and unobserved instances are assumed to be drawn independently from the same distribution. Classification problems are solved using instances' features (content) and labels. Connections/dependencies/ relations between instances are not taken into consideration. On the other hand, learning problems with network information, where for each node its features and relations with other nodes are available, become more common in our lives. Examples include social [39], semantic [36], financial [3], communication [10] and gene regulatory [2] networks. Classification of nodes or links in the

Z. Çataltepe (✉) • A. Sönmez
Istanbul Technical University, Computer Engineering Department, Maslak, Istanbul 34469, Turkey
e-mail: cataltepe@itu.edu.tr; sonmezab@itu.edu.tr

Ş. Gündüz-Öğüdücü and A.Ş. Etaner-Uyar (eds.), *Social Networks: Analysis and Case Studies*, Lecture Notes in Social Networks, DOI 10.1007/978-3-7091-1797-2_7,

network, discovery of links or nodes which are not yet observed or identification of essential nodes or links, are some of the research areas in social networks.

In a social network, the class membership of one node may have an influence on the class membership of a related node. Networked data contain not only the features for each node, but also features and sometimes labels of neighbors and sometimes the features and links of the nodes for which labels need to be estimated. Classification in social networks have a number of challenges compared to classification of plain data:

- Design of classifiers which can make the best use of both content and link information available in the social network data is a challenge.
- One other issue that must be considered is the separation of data into training and test sets, because using traditional random sampling methods may not give healthy results [21].
- Even nodes connected through a chain of links may affect each other's class. Classification algorithms may need to take this dependency into account.

In this chapter, we try to address these challenges. Section 4 describes random and snowball sampling which can be used to partition networked data into training and test sets so that different learning methods can be tested against each other. In Sect. 5 we show how neighbor labels may be used through different aggregation mechanisms. In Sect. 6 we discuss the classification methods for networked data and especially the collective classification algorithms, which allow test nodes to be labeled based on the actual labels of training nodes and the estimated labels of test nodes.

In addition to these, we give the notation used in Sect. 2. Details on graph properties, which are important in understanding social networks data and which algorithms to use for learning, are given in Sect. 3. Experimental results on three different datasets are in Sect. 7. The chapter is concluded with Sect. 8.

2 Notation

We assume that there is a networked dataset represented by a graph $G = (V, E)$ with nodes (vertices) V and undirected links (edges) $E \subseteq \{\{u, v\} | u, v \in V\}$.

Each node $u \in V$ has a C dimensional label vector $\mathbf{r}(u) \in \{0, 1\}^C$ which uses 1-of-K representation and shows the class of the node. Some of the vertices are in the training set V_{train} whose labels are known, while the rest are in the test set V_{test} whose labels will be predicted. Note that, $V_{train} \cap V_{test} = \emptyset$ and $V_{train} \cup V_{test} = V$. Note also that if the test set is unknown then $V_{test} = \emptyset$.

Each node $u \in V$ (whether it is in the training or test set) also has a d dimensional node content feature vector $\mathbf{x}(u) \in \{0, 1\}^d$.

In the pattern recognition scenario that we are interested in, given feature vectors of the training nodes and their labels, $\mathbf{x}(u)$ and $\mathbf{r}(u)$, $u \in V_{train}$, we need to train a mapping (classifier) $g(\mathbf{x}(u)) : \{0, 1\}^d \rightarrow \{0, 1\}^C$. Since the classifier $g(\mathbf{x}(u))$

uses only the input features, we will call it the Content Only (CO) classifier $g(\mathbf{x}(u)) = g_{CO}(\mathbf{x}(u))$.

When not only the training node features, but also their links are given, the link information can also be used for classification. Usually link information of neighbors of a specific node are taken into account. The neighborhood function $N(u)$ returns a set of nodes which are neighbors of node u according to the links L:

$$N(u) = \{v : \{u, v\} \in L\}. \tag{1}$$

Most classifiers need fixed dimensional inputs. So, we need to aggregate [29] labels of neighbors of a node into a fixed dimensional vector. Let $\mathbf{r}_N(u)$ denote the *aggregated neighbor labels* for a node u. We will define $\mathbf{r}_N(u)$ based on the aggregation method (Sect. 5) used.

Based on the labels of the neighbors only, a classifier, which we call the Link Only (LO) classifier $g_{LO}(\mathbf{r}_N(u))$ can be trained on the training data. When a test node needs to be classified, if it has neighbors in test set, inputs to the classifier need to be determined iteratively, based on the current label assignment of the neighbors using a collective classification algorithm such as the Iterative Classification Algorithm (ICA) [21, 34].

When both node features and links are known, a classifier that uses both the content features of the node and labels of the neighbors have been used in [34]. We will call this classifier with $d + C$ features, the Content and Link classifier:

$$g_{CL}([\mathbf{x}(u)\ \mathbf{r}_N(u)]). \tag{2}$$

Evaluation of a classifier can be based on its accuracy on the test set:

$$acc(g, V_{test}) = \frac{1}{|V_{test}|} \sum_{v \in V_{test}} 1 - [g(\mathbf{x}(v)) = \mathbf{r}(v)]. \tag{3}$$

Here $[P]$ is the Iverson bracket and returns 1 if condition P is true (i.e. vectors $g(\mathbf{x}(v))$ and $\mathbf{r}(v)$ are equal) and returns 0 if condition P is false (i.e. $g(\mathbf{x}(v))$ and $\mathbf{r}(v)$ differ in at least one position). The test labels and sometimes the identities of the test inputs are not known during training. Since the actual labels are not known, test accuracy can not be evaluated. Instead, validation set(s) extracted from the labeled training set through sampling (Sect. 4) is(are) used to estimate the test accuracy of a classifier.

3 Graph Properties

Social network data consist of a graph with nodes, node features and labels and links between the nodes. Graph properties, such as homophily, degree distribution help us understand the data at hand. Graph properties can also be used as guidelines in finding out if a sampling (Sect. 4) used for evaluation of algorithms is a good one or not, or which type of aggregation (Sect. 5) of neighbor labels should be used for classification. In this section, we define some of the important graph properties. Please see, for example [27] or [11], for more details on graph properties.

3.1 Homophily

A classification algorithm in a social network aims to use both content and link information. Whether the link information will be useful or not depends on whether linked objects have similar features and or labels. This similarity have been quantified using a number of different criteria.

Neville and Jensen defined two quantitative measures of two common characteristics of relational datasets: *concentrated linkage* and *relational autocorrelation*. Concentrated linkage occurs when many entities are linked to a common entity like the citation links to the key papers. Relational autocorrelation occurs when the features of the entities are similar among entities that share a common neighbor [17]. As pointed out by Neville and Jensen, most of the models (e.g., PRMs, RMNs) do not automatically identify which links are the most relevant to the classification task. In their method, they defined links which are most relevant to the classification task by using concentrated linkage and relational autocorrelation, and explored how to use these relational characteristics to improve feature selection in relational data [42].

Yang et al. identified five hypertext link regularities that might (or not) hold in a particular hypertext corpus [40, 42]. We list three of them here.

- *Encyclopedia regularity:* The class of a document is the same as the class of the majority of the linked documents.
- *Co-referencing regularity:* Documents with the same class tend to link to documents not of that class, but which are topically similar to each other.
- *Partial co-referencing regularity:* Documents with the same class tend to link to documents that are topically similar to each other, but also link to a wide variety of other documents without semantic reason. The presence (or absence) of these regularities may significantly influence the optimal design of a link-based classifier. Most of link analysis methods and link-based classification models are built upon the "encyclopedia" or "co-referencing" regularity. As a result, the models do not automatically identify which links are most relevant to the task [42].

Assortativity index defined by [27] is also a measure of how similar are the labels of connected nodes. Assortativity index is defined using the number of links which connect nodes of same and different classes and it is proportional to the number of links that connect nodes of the same class.

Homophily [21, 31] or label autocorrelation can be defined as the tendency of entities to be related to other similar entities, i.e. linked entities generally have a tendency to belong to the same class. High homophily usually implies that using link information helps with classification while for low homophily datasets using a content only classifier could do a better job.

We define homophily of a node u as the proportion of neighbor nodes of u which have the same label as u:

$$H(u) = \frac{1}{N(u)} \sum_{v \in N(u)} [\mathbf{r}(u) == \mathbf{r}(v)] \tag{4}$$

In this equation [] is the Iverson bracket and $[p]$ is equal to 1 if p is true, it is 0 otherwise. The graph homophily is then:

$$H(G) = \frac{1}{|V|} \sum_{u \in V} H(u) \tag{5}$$

Note that although homophily is defined as label "sameness", labels could be related to each other, and for example could just be the opposite of each other for a binary classification problem (i.e. heterophily). The classifiers that we use would be able to take advantage of homophily, heterophily or any other correlation between the label of a node and labels its neighbors.

3.2 Degree Distribution

Degree, $k(u)$, of a node u is the total number of its connections to the other nodes in the network. Although the degree of a node seems to be a local quantity, degree distribution of the network often may help to determine some important global characteristics of networks. A scale-free network is type of a network whose degree distribution follows a power law [11]. It was shown that whether an epidemic spreads in a scale-free network or stops can be computed based on the scale free exponent in [16].

3.3 Clustering Coefficient

Clustering coefficient is a measure of degree to which nodes tend to cluster together in a graph. Clustering coefficient property can give information, for example, on how close are two nodes in the graph, and therefore how much their predicted labels would affect each other during prediction of labels. Different types of clustering coefficients can be defined as follows.

3.3.1 Global Clustering Coefficient

The global clustering coefficient, which is a measure of indication of the clustering in the whole network is based on triplets of nodes. A triplet is called an open triplet when three nodes are connected with two links and it is called a closed triplet when all three nodes are tied together. Three closed triplets, one centered on each of the nodes, form triangle. The global clustering coefficient is defined as the ratio of the number of closed triplets to the total number of triplets (open and closed) [11]:

$$CC_G(G) = \frac{Number\ of\ closed\ triplets}{Number\ of\ connected\ triplets} \tag{6}$$

3.3.2 Local Clustering Coefficient

Local clustering coefficient is defined in [11] as the ratio of the actual number of edges between the neighbors of a node u to all the possible numbers of edges is the local clustering coefficient for node u:

$$CC_L(u) = \frac{\sum_{v_1,v_2 \in N(u), v_1 \neq v_2} [\{v_1, v_2\} \in E]}{|N(u)|(|N(u)| - 1)/2} \tag{7}$$

Here [] is the Iverson bracket. The number of all possible undirected links between neighbors $N(u)$ of node u is $|N(u)| * (|N(u)| - 1)/2$.

3.3.3 Average Clustering Coefficient

The network average clustering coefficient is defined by Watts and Strogatz as the average of the local clustering coefficients of all the nodes in the graph[38]:

$$CC_L(G) = \frac{1}{|V|} \sum_{u \in V} CC_L(u) \tag{8}$$

If the average clustering coefficient is zero, then the graph is a tree. If the average clustering coefficient is higher than a random graph with the same degree distribution, then the network may show the small world phenomenon, i.e. any two random nodes can be connected using much smaller number of links than $O(|V|)$. It should be noted that during calculation of the average clustering coefficient the nodes having less than two neighbours, which naturally have a clustering coefficient of zero, are not considered in the average clustering coefficient computation.

3.4 Rich Club Coefficients

Let $V_k \subseteq V$ denote the nodes having degree higher than a given value k and $E_{>k}$ denote the edges among nodes in V_k. The rich-club coefficient is defined as:

$$\phi_k(G) = \frac{|E_{>k}|}{|V_{>k}|(|V_{>k}| - 1)/2} \tag{9}$$

In Eq. (9) $|V_{>k}|(|V_{>k}| - 1)/2$ represents the maximum possible number of undirected links among the nodes in $V_{>k}$. Thus, $\phi(k)$ measures the fraction of edges actually exist between those nodes to the maximum number of edges they may have. The rich club coefficient helps to understand important information about the underlying architecture revealing the topological correlations in a complex network [7].

3.5 Degree-Degree Correlation

Degree-degree correlation is the correlation between the number of neighbors (minus 1) of neighboring nodes. Both homophily and degree-degree correlation can be considered as special case of assortativity [26], the correlation between a certain property (label, degree, etc.) of two neighboring nodes. Average degree of the neighbours' of the nodes having degree k. Kahng et al. [19] found out that among scale-free networks authorship and actor networks are assortative (i.e. nodes with large degree connect to nodes with large degree) while protein-protein interaction and world wide web networks are disassortative (i.e. large degree nodes tend to connect to small degree nodes).

3.6 Graph Radius and Diameter

These two notions are related to each other. We use the definitions from [12].

Let $d_G(u, v)$ be the distance (i.e. the minimum number of edges that connect u and v) between nodes u and v. The eccentricity $\epsilon(u)$ of a node u is defined as the greatest distance between u and any other node in the graph:

$$\epsilon(u) = \max_{v \in V} d_G(u, v) \tag{10}$$

The diameter of a graph is defined as the length of the longest of the shortest paths between any two nodes or equivalently as the maximum eccentricity of any vertex:

$$diam(G) = \max_{u,v \in V} d_G(u, v) \tag{11}$$

The radius of the graph is defined as the minimum eccentricity among all the nodes in the graph:

$$rad(G) = \min_{u \in V} \epsilon(u) = \min_{u \in V} \max_{v \in V} d_G(u, v) \tag{12}$$

3.7 Average Path Length

The Average Path Length of a graph is defined as the average of all shortest paths between nodes:

$$APL(G) = avg_{u \in V} \max_{v \in V} d_G(u, v) \tag{13}$$

[13] has computed the average path length based on whether the graph is a scale free one or not and the scale free exponent.

3.8 Graph Density

The density of a graph is defined as the ratio of the number of edges to the number of possible edges:

$$D(G) = \frac{|E|}{|V|(|V| - 1)/2} \tag{14}$$

3.9 Graph Properties of the Datasets

We used three datasets that have been used in network classification research [21, 23, 34]. We give their graph properties in Table 1.

3.9.1 Cora Dataset

Cora [22] data set consists of information on 2708 Machine Learning papers. Every paper in Cora cites or is cited by at least one other paper in the data set. There are 1,433 unique words that are contained at least 10 times in these papers. There are also seven classes assigned to the papers according to their topics. For each paper, whether or not it contains a specific word, which class it belongs to, which papers it cites and which papers it is cited by are known. Citation connections and paper features (class and included words) are contained in two separate files. Total number of connections between the papers is 5,278. There are 3.898 links per paper.

Table 1 Summary information about the datasets (Cora, Citeseer, WebKB)

DataSet	Cora	CiteSeer	WebKB
Number of features	1,433	3,703	1,703
Number of nodes	2,708	3,312	877
Number of unique links	5,278	4,536	1,388
Number of classes	7	6	5
Average degree	3.90	2.74	3.17
Diameter	19	28	8
Average path length	6.31	9.30	3.14
Density	0.0014	0.0008	0.0036
Global clustering coefficient	0.094	0.130	0.0362
Average clustering coefficient	0.293	0.243	0.294
Alpha (power law)	1.867	1.638	1.810
Homophily	0.83	0.71	0.13

3.9.2 Citeseer Dataset

CiteSeer [14, 33] data set consists of information on 3,312 scientific papers. There are 3,703 unique words that are contained at least 10 times in these papers. There are six classes assigned to the papers according to their topics. Just as in the CoRA dataset, word, class and cites and cited by information are given in two separate files. Total number of connections between the papers is 4,536. There are 2.74 links per paper.

3.9.3 WebKB Dataset

WebKB [9] data set consists of sets of web pages from four computer science departments, with each page manually labeled into five categories: course, project, staff, student, or faculty.

Link structure of WebKB is different from Cora and Citeseer since co-citation links are useful for WebKB data set. The reason for that can be explained based on the observation that a student is more likely to have a hyperlink to her adviser or a group/project page rather than to one of her peers [21].

4 Sampling

The test data in social networks may contain a bunch of nodes that are connected to each other, in which case the training-validation partitioning process needs to take this dependency into account. It is also possible that the test data are randomly distributed among the training nodes. Two different sampling mechanisms, snowball sampling and random sampling, are used to handle these two situations.

4.1 Random Sampling

When random sampling is used, nodes in training, validation, and test sets are selected. It is important to preserve class distribution of the dataset as much as possible during selection of the nodes. One method to achieve this is to partition nodes from every class among themselves randomly proportional to the required train, validation and test set sizes and then combine the nodes from every class in the training set to produce the training set and do the same for validation and test sets. While random sampling is a simple method, even if care is taken to preserve the class ratios, the sampled graph is likely to have very different topological properties than the original graph [1], therefore, the classification algorithms trained/tested on the sampled graph may not perform similarly on the test set or the original dataset.

4.2 Snowball Sampling

When the usual k-fold cross-validation training and test sets are obtained on networked data, especially when the number of links per node is low, k-fold random sampling may generate almost disconnected graphs [34], making learning through the links in the networked data impossible. To overcome the issue of disconnected graphs in k-fold cross validation, snowball sampling is used. In snowball sampling, first of all, different starting nodes are selected. Then new nodes are selected among the nodes which are accessible through the selected nodes. Thus, the selected nodes grow like a snowball and as a result selected nodes are connected. It is important to preserve the class ratios in the selected set of nodes, therefore, at every point during sampling, if a class is underrepresented, it is given higher probability. This sampling procedure continues until there are enough nodes in the selected subset.

In Fig. 1, visualization of train/test partitions for Cora and Citeseer datasets are given for both random sampling and snowball sampling. Red nodes are nodes in the test partition while green ones are in the train partition. Nodes' actual labels are also displayed in the circles representing the nodes. In order to be able to visualize these two sets a subsample of 4 % of the actual data is used.

When random sampling is used, there is no order or dependency between the selection of training, validation and test sets. However, when snowball sampling is used, whether the snowball is selected and taken to be the test set or the training set may give different results. Taking the snowball to be the test set [34], generating k disjoint snowballs and using them for training-validation set formation [23], using temporal sampling and past data for training and generating snowball samples with some portion of provided labels for validation [24] are all different uses of snowball sampling for classification of networked data. While some authors mention that snowball sampling causes bias towards highly connected nodes and may be more suitable to infer about links than to infer about nodes in a social network [35], others suggest that since snowball is not guaranteed to reach individuals with

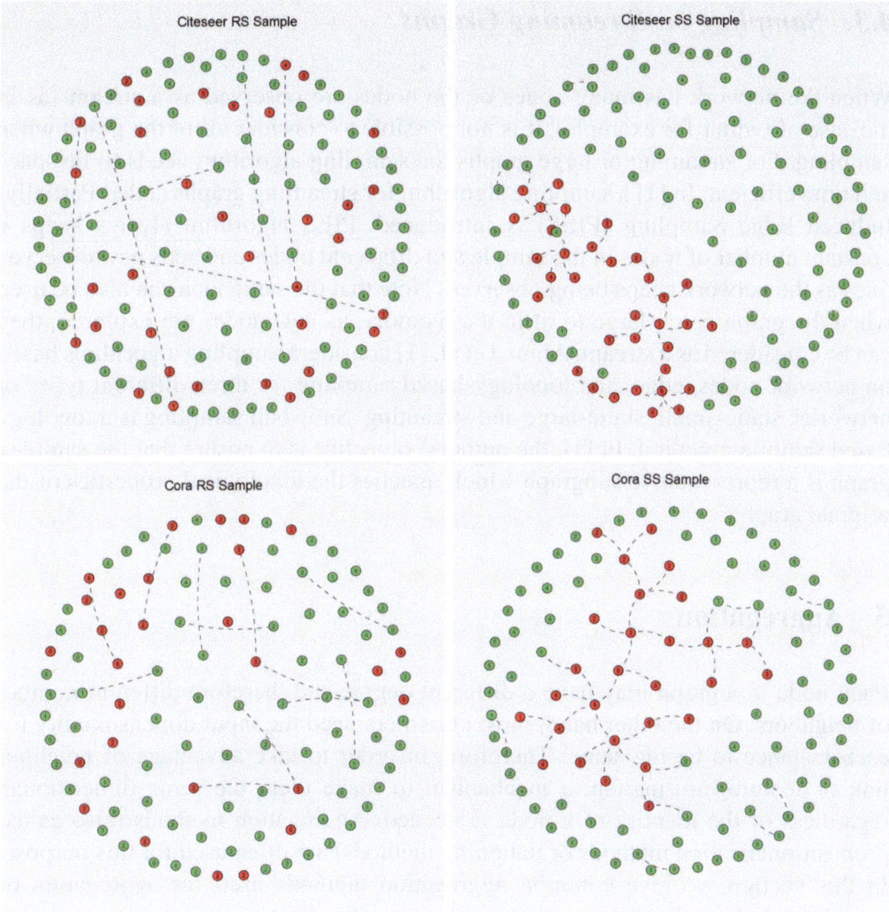

Fig. 1 Random sampling vs snowball sampling

high connectivity and would not reach disconnected individuals, snowball's starting nodes should be chosen carefully [15].

When random sampling is used to generate k-fold cross validation training and validation (and test) sets, there are no overlaps between different test sets. However, when snowball sampling is used to generate the k test sets, the test snowballs created may overlap. Since linked instances in a snowball are correlated, errors made on them may also be correlated. The statistical tests, such as the paired t-test, which is used for model selection, may not give reliable results when test sets are not independent [25]. Forest Fire Sampling (FFS) [20] method may be considered as an alternative to snowball sampling, in this algorithm, as in snowball sampling, a breadth first search technique is used but some of the followed links are burned according to a probability distribution.

4.3 Sampling on Streaming Graphs

When the network has many nodes or the nodes are observed as a stream (as in the case of twitter for example), it is not possible to consider all of the graph when sampling. For streaming or large graphs the sampling algorithm needs to be space and time efficient. In [1] a sampling algorithm for streaming graphs called Partially-Induced Edge Sampling (PIES) is introduced. PIES algorithm always keeps a constant number of nodes in the sample and drops old nodes and adds new observed ones as the network keeps being observed. Note that the same idea can also be used when the graph is too large to fit in the memory, as new nodes are explored, they can be considered as a stream. Ahmed et al. [1] considers sampling algorithms based on network nodes, edge, and topology-based sampling for three different types of networks, static-small, static-large and streaming. Snowball sampling is a topology based sampling method. In [1], the authors' objective is to ensure that the sampled graph is a representative subgraph which matches the topological properties of the original graph.

5 Aggregation

Each node in a graph may have a different degree and therefore different number of neighbors. On the other hand, most classifiers need the input dimensionality for each instance to be the same. Therefore, in order to take advantage of neighbor link or feature information, a mechanism to make them the same dimensional, regardless of the identity of a node is needed. Aggregation methods (also called propositionalization methods or flattening methods) are often used for this purpose. In this section we give common aggregation methods used for aggregation of neighbor labels so that they can be used for classifier training.

The main objective of aggregation in relational modeling is to provide features which improve the generalization performance of the model [29]. However aggregation usually causes loss of information, therefore one needs to be careful about not losing predictive information. Perlich and Provost proposed general guidelines for designing aggregation operators, suggesting that aggregation should be performed keeping the class labels under consideration, aggregated features should cause instances of the same class to be similar to each other and different aggregation operators should be experimented with. Below, we will present performances of different aggregation methods on different datasets. In [29] authors considered both simple aggregators and new more complex aggregators in the context of the relational learning system ACORA (Automated Construction of Relational Attributes). ACORA computes class-conditional distributions of linked object identifiers, and for an instance that needs to be classified, it creates new features by computing distances from these distributions to the values linked to the instance [29]. Lu and Getoor considered various aggregation methods: existence(binary), mode and value counts. The count method performed best in their study [28].

In the following sections, the notation introduced in Sect. 2 is used. Note that, although the neighborhood function $N(u)$ is usually defined to include the immediate neighbors of a node, it could be extended to include neighbors which are at most a number of links away.

5.1 Neighbor Label Aggregation Methods

- Count Method: The count aggregation method [30] determines the frequency of the neighbors having the same class as the node:

$$\mathbf{r}_N^{count}(u) = \sum_{v \in N(u)} \mathbf{r}(v). \tag{15}$$

  The count method does not consider any uncertainty with the labels or links, neither does it consider the edge weights [30].
- Mode Method: This aggregation method considers the mode of the neighbor labels:

$$\mathbf{r}_N^{mode}(u) = mode_{v \in N(u)} \mathbf{r}(v). \tag{16}$$

- Binary Existence Method: This aggregation method only considers whether a certain label exists among the neighbors or not, it does not take into account the number of occurrences, as count or mode aggregation do. For the jth class, the binary existence of neighbor labels' aggregation is computed as:

$$\mathbf{r}_N^{exist}(u, j) = [\mathbf{r}_N^{count}(u, j) > 0] \tag{17}$$

- Weighted Average Method: The weighted average aggregation method [30] sums the weights of the neighbors of the node belonging to each class and then normalizes it with the sum of the weights of all edges to the neighbors. Similar to the count method, it does not consider uncertainty.

$$\mathbf{r}_N^{wavg}(u) = \frac{1}{Z} \sum_{v \in N(u)} w(u, v)\mathbf{r}(v). \tag{18}$$

  Here $w(u, v) \in \mathscr{R}$ is the weight of the link between nodes u and v, Z is a normalization constant:

$$Z = \sum_{v \in N(u)} w(u, v) \tag{19}$$

Table 2 Accuracy comparison of aggregation methods (ICA, SS) (Cora, Citeseer, WebKB)

	Cora	CiteSeer	WebKB
Count method	75.39 ± 1.02	68.86 ± 1.22	84.75 ± 0.67
Mode method	75.42 ± 0.98	69.79 ± 1.00	81.68 ± 0.43
Binary existence method	71.33 ± 1.65	68.58 ± 0.71	80.32 ± 0.86
Proportion method	76.46 ± 1.31	66.81 ± 0.85	76.45 ± 0.93
Pr. weighted average method	64.43 ± 1.15	66.81 ± 0.85	77.82 ± 0.99

Table 3 Accuracy comparison of aggregation methods (ICA,RS) (Cora, Citeseer, WebKB)

	Cora	CiteSeer	WebKB
Count method	87.78 ± 0.48	77.43 ± 0.67	87.01 ± 1.01
Mode method	87.78 ± 0.59	78.13 ± 0.66	85.17 ± 1.93
Binary existence method	85.19 ± 0.61	77.58 ± 0.78	86.78 ± 1.11
Proportion method	86.48 ± 0.63	77.76 ± 0.89	86.32 ± 1.17
Pr. weighted average method	70.15 ± 0.74	70.76 ± 0.93	86.32 ± 1.04

- Probabilistic Weighted Average Method: This aggregation method is the probabilistic version of the Weighted Average method. It is based on the weighted arithmetic mean of class membership probabilities of neighbors of a node. This method was introduced by Macskassy and Provost and was used as a probabilistic Relational Classifier (PRN) classifier [30].

$$\mathbf{r}_N^{pwavg}(u,c) = \frac{1}{Z} \sum_{v \in N(u), c \in C} w(u,v).P(c|v) \tag{20}$$

  where Z is defined as in Eq. (19) and c denotes a certain class.

In Tables 2 and 3, we show the average test accuracies over ten folds, of using iterative classification algorithm (ICA), which is a method for transductive classification of networked data (see the next section). We used logistic regression as the base classifier, during these experiments. As it can be seen from both tables, the count method outperforms the other methods both in terms of its simplicity and the accuracies obtained.

6 Classification in Social Networks

We consider two types of classification in a social network. The content only classification can be used whether the test nodes are known or not. On the other hand, when the test nodes or some unlabeled nodes are known, then semi-supervised classification algorithms can be used.

6.1 *Supervised, Content Only Classification*

This model consists of a (learned) model, which uses only the local features of the nodes whose class label will be estimated. The local models can also be used to generate priors for the initial state for the relational learning and collective inference components. They also can be used as one source of evidence during collective inference. These models typically are produced by traditional machine learning methods [21].

6.2 *Semi-supervised and Transductive Classification in Social Networks*

Since social network data usually come in huge sizes, in addition to labeled instances, there are, usually, a huge number of unlabeled instances. In such cases, it could be possible to use the information other than labels that exist in the unlabeled data, which leads to use of semi-supervised learning algorithms. When the test nodes whose class will need to be predicted are known, then we have a transductive learning scenario.

In contrast to the non-relational (local) model, the relational model use the relations in the network as well as the values of attributes of related entities, even possibly long chains of relations. In relational models, a relational classifier determines the class label or estimates the class conditional probabilities. The relational classifier might combine local features and the labels of neighbors using a naive Bayes model or a logistic regression [21].

In semi-supervised learning [43], the unlabeled instances can be used to monitor the variance of the produced classifiers, to maximize the margin and hence to minimize the complexity [18], to place classifier boundaries around the low density regions between clusters in the data [6]. There are also co-training [5] type algorithms which need different classifiers that are obtained through the use of different type of classifiers, different feature subspaces [41] or set of instances. When the classifiers produced are diverse and accurate enough, co-training may improve the final test accuracy [37]. On the other hand, Cozman and Cohen [8] have shown that unlabeled data can degrade the classification performance when there are discrepancies between modeling assumptions used to build the classifier and the actual model that generates the data. Therefore, both for the general semi-supervised and the transductive learning, the use of unlabeled data is not guaranteed to improve performance.

6.3 *Collective Classification*

Collective classification methods, which are sometimes also called collective inference methods, are iterative procedures, which classify related instances simultaneously [30,42]. In collective classification, the content and link information for both training and test data are available. First, based on the available training content, link and label information, models are trained. Then, those models are used to label the test data simultaneously and iteratively where each test sample is labeled based on its neighbors.

Collective classification exploits relational autocorrelation. Relational autocorrelation is a very important property of relational data and is used as a measure of how an attribute for an instance is correlated with the same variable from a related instance [30].

However, sometimes, the advantage of exploiting the relationships can become a disadvantage since it is possible to make incorrect predictions about a particular node which propagates in the network and may lead to incorrect predictions about other nodes. Bilgic and Getoor proposed an acquisition method which learns the cases when a given collective classification algorithm makes mistakes, and suggests label acquisitions to correct those mistakes [4].

Iterative classification algorithm (ICA) and Gibbs sampling algorithm (GS), Mean field relaxation labeling (MF), Loopy belief propagation (LBP), are popular approximate inference algorithms used for collective classification [34]. In this book chapter, we explain the ICA algorithm and use it in our experiments. Iterative classification algorithm (ICA) is a popular and simple approximate collective inference algorithm [21, 34]. Despite its simplicity, ICA was shown to perform as well as the other algorithms such as Gibbs Sampling [33]. Please see [21, 34] for details on the other collective classification algorithms.

6.3.1 Iterative Classification Algorithm (ICA)

To determine the label of a node, Iterative Classification Algorithm (ICA) assumes that all of the neighbors' attributes and labels of that node are already known. Then, it calculates the most likely label with a local classifier which uses node content and neighbors' labels. However, most nodes have neighbors which are not in training data and hence are not labeled, therefore the label assignment on one test instance may affect the label assignment on a related test instance. ICA repeats the labeling process iteratively until all of the label assignments are stabilized. Neighbor label information is summarized using an aggregation operator (Sect. 5.1).

Pseudocode for the ICA algorithm (based on [34]) is given in Algorithm 1. In the pseudo code, $\tilde{\mathbf{r}}(u)$ stands for temporary label assignment of instance u in the test set. $g_{CL}([\mathbf{x}(u)\ \mathbf{r}_N(u)])$ is the base classifier which is first trained on training nodes and their neighbors from the training set. The base classifier uses the estimated labels of the neighbors if they are test nodes. O is a random ordering of test nodes.

As shown above, the ICA algorithm starts with a bootstrapping to assign initial temporary labels to all nodes by using only the content features of the nodes. Then, it starts iterating and updating labels according to the both relational and content features [32].

Algorithm 1: $\tilde{\mathbf{r}}(V_{test}) = \text{ICA}(G, V_{train}, V_{test}, g_{CL}())$

for all $u \in V_{test}$ **do**
  Compute $\tilde{\mathbf{r}}_N(u)$ using only neighbors in V_{train}
  Set $\tilde{\mathbf{r}}(u) \leftarrow g_{CL}([\mathbf{x}(u)\ \tilde{\mathbf{r}}_N(u)])$
end for
repeat
  Generate ordering O over nodes in V_{test}
  for all $u \in O$ **do**
    Compute $\tilde{\mathbf{r}}_N(u)$ using current label assignments to nodes in $N(u)$
    Set $\tilde{\mathbf{r}}(u) \leftarrow g_{CL}([\mathbf{x}(u)\ \tilde{\mathbf{r}}_N(u)])$
  end for
until all labels are stabilized or threshold number of iterations

7 Experiments

In this section, we report classification results on three networked datasets. We evaluate content only (CO), ICA using only a link based classifier (LO), ICA using a content and link based classifier (ICA). We show that whether a method works on a dataset or not is highly dependent on the properties, especially homophily of the dataset. In the experiments, we report results using both snowball and random sampling methods to select the test set. We show that sampling may also affect the results.

7.1 Experimental Setup

We use Cora, Citeseer and WebKb datasets whose details are given in Table 1. Both Cora and Citeseer datasets consist of information on scientific papers. As features, the words that occur at least ten times are used. For each paper, whether or not it contains a specific word, which class it belongs to, which papers it cites and which papers it is cited by are known. The WebKB dataset contains web pages, instead of papers as content. The hyperlinks between web pages are used to produce links between nodes.

We experiment with different classification methods, namely, logistic regression (LR), Support Vector Machine (SVM), Naive Bayes (NB), Bayes Net (BN), k-Nearest Neighbor (kNN, k = 3) for the g_{CO}, g_{LO} and g_{CL} classifiers. For all of the methods Weka implementations with default parameters (unless otherwise noted) have been used.

7.2 Performance of Different Classifiers

First of all, we conducted experiments on datasets using Logistic Regression (LR), Support Vector Machine (SVM), Naive Bayes (NB), Bayes Net (BN), k-Nearest

Table 4 Accuracies on Cora dataset using different classifiers (snowball sampling)

	Acc(CO)	Acc(LO)	Acc(ICA)
Logistic regression	0.61 ± 0.01	0.77 ± 0.01	0.72 ± 0.01
SVM	0.70 ± 0.01	0.64 ± 0.02	0.71 ± 0.01
KNN(k = 1)	0.46 ± 0.01	0.66 ± 0.02	0.52 ± 0.01
KNN(k = 3)	0.47 ± 0.01	0.65 ± 0.01	0.54 ± 0.01
Naive Bayes	0.70 ± 0.01	0.60 ± 0.01	0.76 ± 0.01
Bayes Net	0.70 ± 0.01	0.63 ± 0.02	0.78 ± 0.01
Random forest	0.59 ± 0.01	0.61 ± 0.02	0.66 ± 0.01
J48	0.64 ± 0.01	0.52 ± 0.02	0.51 ± 0.03
LibSVM	0.29 ± 0.01	0.61 ± 0.02	0.72 ± 0.01

Table 5 Accuracies on WebKb dataset using different classifiers (snowball sampling)

	A.(CO)	A.(LO)	A.(ICA)
Logistic regression	0.67 ± 0.04	0.56 ± 0.02	0.59 ± 0.05
SVM	0.83 ± 0.02	0.52 ± 0.04	0.83 ± 0.02
KNN1	0.63 ± 0.02	0.46 ± 0.03	0.64 ± 0.02
KNN3	0.56 ± 0.02	0.45 ± 0.02	0.56 ± 0.02
Naive Bayes	0.75 ± 0.02	0.47 ± 0.02	0.75 ± 0.02
Bayes Net	0.78 ± 0.02	0.50 ± 0.02	0.79 ± 0.02
Random forest	0.68 ± 0.02	0.48 ± 0.02	0.70 ± 0.01
J48	0.69 ± 0.02	0.43 ± 0.03	0.67 ± 0.03
LibSVM	0.60 ± 0.02	0.51 ± 0.02	0.64 ± 0.02

Table 6 Accuracies on Citeseer dataset using different classifiers (snowball sampling)

	Acc(CO)	Acc(LO)	Acc(ICA)
Logistic regression	0.55 ± 0.01	0.59 ± 0.02	0.57 ± 0.01
SVM	0.70 ± 0.01	0.56 ± 0.01	0.71 ± 0.01
KNN(k = 1)	0.42 ± 0.01	0.54 ± 0.01	0.43 ± 0.01
KNN(k = 3)	0.38 ± 0.01	0.54 ± 0.02	0.35 ± 0.01
Naive Bayes	0.69 ± 0.01	0.54 ± 0.02	0.73 ± 0.01
Bayes Net	0.70 ± 0.00	0.44 ± 0.03	0.72 ± 0.01
Random forest	0.57 ± 0.01	0.45 ± 0.02	0.57 ± 0.02
J48	0.59 ± 0.00	0.38 ± 0.02	0.44 ± 0.02
LibSVM	0.19 ± 0.00	0.49 ± 0.01	0.60 ± 0.02

Neighbor (kNN) classifiers. We evaluated the accuracies obtained when content only (CO), link only (LO) classifiers and ICA are used with a specific classification method for each dataset.

Tables 4, 5, and 6 show the accuracies obtained when different classifiers are used on Cora CiteSeer and WebKb datasets when Snowball Sampling is used.

Tables 7, 8, and 9 show the accuracies obtained when different classifiers are used on Cora CiteSeer, and WebKb datasets when Random Sampling is used.

Table 7 Accuracies on Citeseer dataset using different classifiers (random sampling)

	Acc(CO)	Acc(LO)	Acc(ICA)
Logistic regression	0.58 ± 0.02	0.68 ± 0.02	0.61 ± 0.02
SVM	0.75 ± 0.01	0.66 ± 0.02	0.75 ± 0.01
KNN(k = 1)	0.42 ± 0.02	0.66 ± 0.02	0.45 ± 0.02
KNN(k = 3)	0.35 ± 0.02	0.65 ± 0.02	0.37 ± 0.02
Naive Bayes	0.71 ± 0.01	0.59 ± 0.03	0.74 ± 0.01
Bayes Net	0.71 ± 0.01	0.65 ± 0.02	0.77 ± 0.01
Random forest	0.60 ± 0.01	0.64 ± 0.02	0.65 ± 0.01
J48	0.63 ± 0.01	0.61 ± 0.02	0.60 ± 0.02
LibSVM	0.16 ± 0.02	0.66 ± 0.02	0.62 ± 0.02

Table 8 Accuracies on Cora dataset evaluated using different classifiers (random sampling)

	Acc(CO)	Acc(LO)	Acc(ICA)
Logistic regression	0.63 ± 0.01	0.85 ± 0.01	0.72 ± 0.01
SVM	0.73 ± 0.01	0.64 ± 0.02	0.76 ± 0.01
KNN(k = 1)	0.48 ± 0.01	0.80 ± 0.01	0.55 ± 0.01
KNN(k = 3)	0.50 ± 0.01	0.81 ± 0.01	0.57 ± 0.01
Naive Bayes	0.73 ± 0.01	0.70 ± 0.02	0.80 ± 0.01
Bayes Net	0.73 ± 0.01	0.78 ± 0.01	0.86 ± 0.01
Random forest	0.64 ± 0.01	0.77 ± 0.01	0.77 ± 0.01
J48	0.65 ± 0.01	0.73 ± 0.01	0.70 ± 0.01
LibSVM	0.33 ± 0.01	0.80 ± 0.01	0.76 ± 0.01

Table 9 Test accuracy results of WebKb dataset evaluated using different classifiers (random sampling)

	A.(CO)	A.(LO)	A.(ICA)
Logistic regression	0.65 ± 0.05	0.52 ± 0.03	0.55 ± 0.05
SVM	0.81 ± 0.03	0.50 ± 0.03	0.81 ± 0.03
KNN1	0.52 ± 0.04	0.48 ± 0.03	0.52 ± 0.04
KNN3	0.51 ± 0.05	0.47 ± 0.03	0.52 ± 0.05
Naive Bayes	0.74 ± 0.03	0.46 ± 0.05	0.74 ± 0.03
Bayes Net	0.77 ± 0.02	0.50 ± 0.03	0.80 ± 0.02
Random forest	0.67 ± 0.03	0.47 ± 0.03	0.69 ± 0.03
J48	0.68 ± 0.03	0.46 ± 0.03	0.67 ± 0.03
LibSVM	0.58 ± 0.05	0.50 ± 0.03	0.62 ± 0.04

For CO classification, while LR, SVM, BN and NB give similar accuracies, kNN usually gives worse accuracies. On the other hand, for LO all classifiers perform similarly.

For the CO classification, SVM, NB and BN classifiers usually performed better than the others. We think that this is due to the high input dimensionality of the content features. On the other hand, for LO classification LR outperformed the other methods. Since the LO homophily of Cora dataset is higher than the CiteSeer, the LO accuracies are also higher. For the Cora dataset, instead of using thousands of features in CO classifier, simply using the aggregated class neighbors in LO classifier results in a better accuracy. This is expected since both datasets have high homophily and LO (and ICA) benefits from homophily. For both Cora and Citeseer datasets, BN method gives the best results for ICA and ICA performs better than CO methods. However, again due to high link graph homophily, ICA performs just a little better than LO accuracies.

8 Conclusion

In this chapter, we have given details on how content, link and label information on social network data can be used for classification. We defined important properties of social networked data, which may be used to characterize a networked dataset. We defined a list of aggregation operators, which are used to aggregate the labels of neighbors, so that the classifiers trained have the same number of inputs for each node. We evaluated the accuracies of different classifiers, using only the content, only the link or both the content and the link information.

We have shown that graph properties, especially homophily, play an important role on whether network information would help in classification accuracy or not. The Cora dataset which has high homophily benefits a lot from the use of network information. It is also worthwhile noting that when homophily is high, one can only use the link information and may not need the content information at all.

We experimented with a number of label aggregation methods and found out that the count method, which is one of the simplest aggregation methods, is as good as the other methods.

The networked data does not obey the i.i.d. assumptions which are mostly assumed to hold for content only datasets. The test dataset also may come either randomly among distributed in the network, or they may be concentrated on certain parts of the network. We used random and snowball sampling algorithms to separate the a portion of the all available data as test data. We have shown that, with random sampling, the test accuracies of classifiers that use network information are always better. This is due to the fact that with random partition the nodes in the test partition will naturally have more neighbors from train and validation partitions, which have the actual labels, as opposed to the labels estimated by the classifiers.

Depending on whether content only, link only or content and link information is used and depending on the dataset and the type of content information, we have shown that different types of classifiers, such as SVM, Naive Bayes, Logistic Regression classifier, may perform differently. All the datasets we used in our experiments contained text as content information, therefore, classifiers which were able to deal with high dimensional features, such as SVM, Naive Bayes and Bayes Net performed better for content only or content and link classification. On the other hand, when only link information was used, the logistic regression classifier performed better. We suggest that using different type of classifiers and content only, link only and content and link classification should be experimented with for a networked dataset.

References

1. Ahmed NK, Neville J, Kompella R (2012) Network sampling: from static to streaming graphs. arXiv preprint arXiv:1211.3412
2. Awan A, Bari H, Yan F, Moksong S, Yang S, Chowdhury S, Cui Q, Yu Z, Purisima EO, Wang E (2007) Regulatory network motifs and hotspots of cancer genes in a mammalian cellular signalling network. IET Syst Biol 1(5):292–297
3. Bernstein AA, Clearwater S, Hill S, Perlich C, Provost F (2002) Discovering knowledge from relational data extracted from business news. In: Proceedings of the workshop on multi-relational data mining at KDD-2002, pp 7–22
4. Bilgic M, Getoor L (2008) Effective label acquisition for collective classification. In: Li Y, Liu B, Sarawagi S (eds) Proceedings of the 14th ACM SIGKDD international conference on knowledge discovery and data mining, Las Vegas, 24–27 August 2008. ACM, New York, pp 43–51
5. Blum A, Mitchell T (1998) Combining labeled and unlabeled data with co-training. In: COLT, pp 92–100
6. Chapelle O, Zien A (2005) Semi-supervised classification by low density separation. In: AISTAT 2005
7. Colizza V, Flammini A, Serrano MA, Vespignani A (2006) Detecting rich-club ordering in complex networks. Nat Phys 2(2):110–115
8. Cozman FG, Cohen I (2002) Unlabeled data can degrade classification performance of generative classifiers. In: FLAIRS 2002
9. Cravenand M, DiPasquo D, Freitag D, McCallum A, Mitchell T, Nigam K, Slattery S (1998) Learning to extract symbolic knowledge from the world wide web. In: Proceedings of the fifteenth national conference on artificial intelligence (AAAI), Madison, pp 509–516
10. Dasgupta K, Singh R, Viswanathan B, Chakraborty D, Mukherjea S, Nanavati AA, Joshi A (2008) Social ties and their relevance to churn in mobile telecom networks. In: EDBT08
11. Dorogovtsev SN, Mendes JF (2002). Evolution of networks. Adv Phys 51(4):1079–1187
12. Erdos P, Pach J, Pollack R, Tuza Z (1989) Radius, diameter, and minimum degree. J Comb Theory Ser B 47(1):73–39
13. Fronczak A, Fronczak P, Holyst JA (2004) Average path length in uncorrelated random networks with hidden variables. Phys Rev E 70(5):056110, 1–7
14. Giles CL, Bollacker K, Lawrence S (1998) Citeseer: an automatic citation indexing system. In: ACM digital libraries, pp 89–98
15. Hanneman RA, Riddle M (2005) Introduction to social network methods. University of California, Riverside
16. Hein O, Schwind M, Konig W (2006) Scale-free networks: the impact of fat tailed degree distribution on diffusion and communication processes. Wirtschaftsinformatik 48(4):267–275
17. Jensen D, Neville J (2002) Linkage and autocorrelation cause feature selection bias in relational learning. In: Proceedings of the 19th international conference on machine learning. Morgan Kaufmann, San Francisco, pp 259–266
18. Joachims T (2003) Transductive learning via spectral graph partitioning. In: ICML 2003, pp 200–209
19. Kahng GO, Oh E, Kahng B, Kim D (2003) Betweenness centrality correlation in social networks. Phys Rev E 67:01710–01711

20. Leskovec J, Faloutsos C (2006, August) Sampling from large graphs. In: Proceedings of the 12th ACM SIGKDD international conference on Knowledge discovery and data mining, ACM, pp 631–636
21. Macskassy SA, Provost F (2007) Classification in networked data: a toolkit and a univariate case study. J Mach Learn Res 8:935–983
22. McCallum A, Nigam K, Rennie J, Seymore K (2000) Automating the construction of internet portals with machine learning. Inf Retr J 3(2):127–163
23. McDowell L, Gupta KM, Aha DW (2007) Cautious inference in collective classification. In: AAAI. AAAI Press, Menlo Park, pp 596–601
24. Neville J, Jensen D (2008) A bias/variance decomposition for models using collective inference. Mach Learn 73(1):87–106
25. Neville J, Gallagher B, Eliassi-Rad T (2009) Evaluating statistical tests for within-network classifiers of relational data. In: ICDM
26. Newman MEJ (2003) Mixing patterns in networks. Phys Rev E 67(2):026126
27. Newman MEJ (2003) The structure and function of complex networks. SIAM Rev 45(2):167256
28. Perlich C, Huang Z (2005) Relational learning for customer relationship management. In: Proceedings of international workshop on customer relationship management: data mining meets marketing
29. Perlich C, Provost FJ (2006) Distribution-based aggregation for relational learning with identifier attributes. Mach Learn 62(1–2):65–105
30. Preisach C, Schmidt-Thieme L (2008) Ensembles of relational classifiers. Knowl Inf Syst 14(3):249–272
31. Provost F, Perlich C, Macskassy S (2003) Relational learning problems and simple models. In: Proceedings of the IJCAI-2003 workshop on learning statistical models from relational data
32. Sen P, Getoor L (2006) Empirical comparison of approximate inference algorithms for networked data. In: ICML workshop on open problems in statistical relational learning (SRL2006)
33. Sen P, Getoor L (2007) Link-based classification. UM Computer Science Department, Technical Report, CS-TR-4858. University of Maryland
34. Sen P, Namata G, Bilgic M, Getoor L, Gallagher B, Eliassi-Rad T (2008) Collective classification in network data. AI Mag 29(3):93–106
35. Snijders TAB (1992) Estimation on the basis of snowball samples: How to weight? BMS: Bull Sociol Methodol (Bulletin de M thodologie Sociologique) 36(1):59–70
36. Tresp V, Bundschus M, Rettinger A, Huang Y (2008) Towards machine learning on the semantic web. In: Uncertainty reasoning for the semantic web I. Lecture notes in AI. Springer, Berlin
37. Wang W, Zhou ZH (2007) Analyzing co-training style algorithms. In: *Machine Learning: ECML 2007* (pp. 454–465). Springer Berlin Heidelberg
38. Watts DJ, Strogatz SH (1998) Collective dynamics of 'small-world' networks. Nature 393: 440–442
39. Xiang R, Neville J, Rogati M (2010) Modeling relationship strength in online social networks. In: Proceedings of the 19th international conference on World wide web. ACM, New York, pp 981–990
40. Yang Y, Slattery S, Ghani R (2002) A study of approaches to hypertext categorization. J Intell Inf Syst 18(2/3):219–241
41. Yaslan Y, Cataltepe Z (2010) Co-training with relevant random subspaces. Neurocomputing 73(10–12):1652–1661
42. Yonghong T, Tiejun H, Wen G (2006) Latent linkage semantic kernels for collective classification of link data. J Intell Inf Syst 26(3):269–301
43. Zhu X (2007) Semi-supervised learning literature survey. Computer Sciences Technical Report, TR 1530. University of Wisconsin Madison

Experiences Using BDS: A Crawler for Social Internetworking Scenarios

Francesco Buccafurri, Gianluca Lax, Antonino Nocera, and Domenico Ursino

Abstract In new generation social networks, we expect that the paradigm of Social Internetworking Scenarios (SISs) will be more and more important. In this new scenario, the role of Social Network Analysis is of course still crucial but the preliminary step to do is designing a good way to crawl the underlying graph. While this aspect has been deeply investigated in the field of social networks, it is an open issue when moving towards SISs. Indeed, we cannot expect that a crawling strategy, good for social networks, is still valid in a Social Internetworking scenario, due to the specific topological features of this scenario. In this paper, we first confirm the above claim and then, define a new crawling strategy specifically conceived for SISs, which overcomes the drawbacks of the state-of-the-art crawling strategies. After this, we exploit this crawling strategy to investigate SISs to understand their main properties and features of their main actors (i.e., bridges).

1 Introduction

In recent years, (online) social networks (OSN, for short) have become one of the most popular communication media on the Internet [31]. The resulting universe is a constellation of several social networks, each forming a community with specific connotations, also reflecting multiple aspects of people personal life. Despite this inherent heterogeneity, the possible interaction among distinct social networks is the basis of a new emergent internetworking scenario enabling a lot of strategic applications whose main strength will be just the integration of possibly different communities yet preserving their diversity and autonomy. This concept is very recent and only a few commercial attempts to implement Social Internetworking

F. Buccafurri (✉) • G. Lax • A. Nocera • D. Ursino
DIIES, University Mediterranea of Reggio Calabria Via Graziella,
Località Feo di Vito, 89122 Reggio Calabria, Italy
e-mail: bucca@unirc.it; lax@unirc.it; a.nocera@unirc.it; ursino@unirc.it

Ş. Gündüz-Öğüdücü and A.Ş. Etaner-Uyar (eds.), *Social Networks: Analysis and Case Studies*, Lecture Notes in Social Networks, DOI 10.1007/978-3-7091-1797-2_8,

Scenarios (SISs, for short) have been proposed [9, 10, 21, 22, 24, 51]. In this new scenario, the role of Social Network Analysis [4, 15, 34, 44, 53, 57, 62] is of course still crucial in studying the evolution of structures, individuals, interactions, and so on, and in extracting powerful knowledge from them. But an important prerequisite is to have a good way to crawl the underlying graph. In the past, several crawling strategies for single social networks have been proposed. Among them, the most representative ones are Breadth First Search (BFS, for short) [62], Random Walk (RW, for short) [41] and Metropolis-Hastings Random Walk (MH, for short) [27]. They were largely investigated for single social networks highlighting their pros and cons [27, 36]. But, what happens when we move towards Social Internetworking Scenarios? In fact, the question opens a new issue that, to the best our knowledge, has not been investigated in the literature. Indeed, this issue is far from being trivial, because we cannot expect that a crawling strategy, good for social networks, is still valid in a Social Internetworking Scenario, due to the specific topological features of this scenario.

This paper gives a contribution in this setting. In particular, through a deep experimental analysis of the above existing crawling strategies, conducted in a multi-social-network setting, it reaches the conclusion that they are little adequate to this new context, enforcing the need of designing new crawling strategies specific for SISs. Starting from this result, this paper gives a second important contribution, consisting in the definition of a new crawling strategy, called *Bridge-Driven Search* (BDS, for short), which relies on a feature strongly characterizing a SIS. Indeed BDS is centered on the concept of *bridge*, which represents the structural element that interconnects different social networks. Bridges are those nodes of the graph corresponding to users who joined more than one social network and explicitly declared their different accounts. By an experimental analysis we show that BDS fits the desired features, overcoming the drawbacks of existing strategies.

As a third important contribution, with the support of such a crawler specifically designed for SISs, we extract data from SISs to detect the main properties of this new kind of scenario and, especially of its main actors, which are bridges. The analysis of bridges, aiming at estimating both classical Social Network Analysis parameters and new specific ones, is conducted in such a way as to discover the nature of bridges in a very deep fashion. For this purpose, a large number of experiments is performed, to derive knowledge about the following topics:

- distribution of the contact number of bridges (hereafter, bridge degree) and non-bridges;
- correlation between bridges and power users (i.e., nodes having a very high degree, generally higher than the average degree of the social network joined by them);
- existence of preferential ties among bridges;
- centrality of bridges in a SIS and in its single social networks.

The results of our analysis provide knowledge about these topics with a strong experimental support and discover even unexpected conclusions about bridges and, in general, a complete knowledge of these crucial elements of Social Internetworking Scenarios.

The plan of this paper is as follows: in Sect. 2, we present related literature. In Sect. 3, we illustrate and validate our Bridge Driven Search approach. In Sect. 4, we describe our experiences devoted to define the main features of SISs and of bridges. Finally, in Sect. 5, we draw our conclusions and we present possible future issues in this research field.

2 Related Literature

In this section, we survey the scientific literature related to our paper. In particular, we first describe the most known techniques proposed to crawl social networks and then we focus on the approaches proposed for Social Network Analysis.

Concerning the former issue, we observe that with the increase in both the number and the dimension of social networks, the development of approaches to sample social networks has become a very challenging issue. The problem of sampling from large graphs is discussed in [38]. In this paper, the authors aim at answering questions such as: (1) which sampling method to use; (2) how small can the sample size be; (3) how to scale up the measurements of the sample to get estimates for larger graphs; (4) how success can be measured. In their activity they consider several sampling methods and check the goodness of their sampling strategies on several datasets.

A technique based on both sampling and the randomized notion of *focus* is proposed in [52]. This method stores samples in a relational database and favors the visualization of massive networks. In this work, the authors specify features frequently characterizing massive networks and analyze the conditions allowing their preservation during the sampling task. An investigation of the statistical properties of sampled scale-free networks is proposed in [37]. In this paper, the authors present three sampling methods, analyze the topological properties of obtained samples, and compare them with those of the original network. Furthermore, they explain the reasons of some emerged biased estimations and provide suitable criteria to counterbalance them.

Methods to produce a small realistic sample from a large real network are presented in [33]. Here, the authors show that some of the proposed methods maintain the key properties of the initial graph even with a sample size down to 30 %. In [62], the social network graph crawling problem is investigated in such a way as to answer questions such as: (1) how fast crawlers into consideration discover nodes/links; (2) how different social networks and the number of protected users affect crawlers; (3) how major graph properties are studied. All these investigations are performed by analyzing samples derived from four social networks, i.e. Flickr, LiveJournal, Orkut and YouTube.

A framework of parallel crawlers based on BFS and operating on eBay is described in [14]. This framework exploits a centralized queue. The crawlers operate independently from each other so that the failure of one of them does not influence the others. In spite of this, no redundant crawling occurs. In [36], the impact of different graph traversal techniques (namely, BFS, DFS, Forest Fire and Snowball Sampling) on the computation of the average node degree of a network is analyzed. In particular, the authors quantify the bias of BFS in estimating the node degree w.r.t. the fraction of sampled nodes. Furthermore, they show how this bias can be corrected. An analysis of the Facebook friendship graph is proposed in [27]. In this activity, the authors examine and compare several candidate crawling strategies, namely BFS, Random Walk, Metropolis-Hastings Random Walk and Re-Weighted Random Walk. They investigate also diagnostics to assess the quality of the samples obtained during the data collection process.

Concerning the main difference between the new crawler BDS proposed in our paper and the above crawling techniques, we highlight the fundamental difference is that the above techniques are not specifically designed to operate effectively on a SIS. This will be confirmed by the experimental analysis provided in Sect. 3.2.

As far as the latter issue dealt with in this section (i.e., Social Network Analysis) is concerned, we observe that studies on Social Networks attracted mainly sociologists. For instance, [58] introduced the six-degrees of separation and the small-world theories. The effects of these theories are analyzed in [19]. Granovetter [28] showed that a Social Network can be partitioned into "strong" and "weak" ties, and that strong ties are tightly clustered. In a second time, with the development of OSNs, Social Network Analysis attracted computer scientists and many studies have been proposed, which investigate the features of one OSN or compare more OSNs. Most of them collect data from one or more OSNs, map these data onto graphs and analyze their structural properties. These approaches are based on the observation that topological properties of graphs may be reliable indicators of the behaviors of the corresponding users [31].

Studies about how an attacker discovers a social graph can be found in [7, 32]. The sole purpose of the attacker is to maximize the number of nodes/links that can be discovered. As a consequence, these two papers do not examine other issues, such as biases.

In [2], the authors compare the structures of Cyworld, MySpace and Orkut. In particular, they analyze the degree distribution, the clustering property, the degree correlation and the evolution over time of Cyworld. After this, they use Cyworld to evaluate the snowball sampling method exploited to sample MySpace and Orkut. Finally, they perform several interesting analyses on the three social networks.

Given a communication network, the approach of [26] aims at recognizing the network topology and at identifying important nodes and links in it. Furthermore, it proposes several compression schemes exploiting auxiliary and purely topological information. Finally, it examines the properties of such schemes and analyzes what structural graph properties they preserve when applied to both synthetic and real-world networks.

In [43], the authors present a deep investigation of the structure of multiple OSNs. For this purpose, they examine data derived from four popular OSNs, namely Flickr, YouTube, LiveJournal and Orkut. Crawled data regard publicly accessible user links on each site. Obtained results confirm the power law, small-world and scale-free properties of OSNs and show that these contain a densely connected core of high-degree nodes.

In [35], the authors focus on analyzing the giant component of a graph. Moreover, they define a generative model to describe the evolution of the network. Finally, they introduce techniques to verify the reliability of this model. In [3], the authors investigate the main features of groups in LiveJournal and propose models that represent the growth of user groups over time. In [40], data crawled from LiveJournal are examined to investigate the possible correlations between friendship and geographic location in OSNs. Moreover, the authors show that this correlation is strong. Carrington et al. [12] proposes a methodology to discover possible aggregations of nodes covering specific positions in a graph (e.g., central nodes), as well as very relevant clusters. Still on clustering, De Meo et al. [18] recently proposed an efficient community detection algorithm, particularly suited for OSNs, and tested its performance against a large sample of Facebook (among other OSN samples), observing the emergence of a strong community structure. In [50], the authors propose *Social Action*, a system based on attribute ranking and coordinated views to help users to systematically examine numerous Social Network Analysis measures. In [13], the authors present an analysis of Facebook devoted to investigate the friendship relationships in this OSN. To this purpose, they examine the topological properties of graphs representing data crawled from this OSN by exploiting two crawling strategies, namely BFS and Uniform Sampling. A further analysis of Facebook can be found in [59]. In this paper, the authors crawled Facebook by means of BFS and formalized some properties such as assortativity and interaction. These can be verified in small regions but cannot be generalized to the whole graph.

Monclar et al. [45], Ghosh and Lerman [23], Onnela and Reed-Tsochas [49], and Romero et al. [54] present approaches for the identification of influential users, i.e. users capable of stimulating others to join OSN activities and/or to actively operate in them. In [1, 39, 55], the authors suitably model the blogosphere to perform leader identification. In [42], the authors first introduce the concept of starters (i.e., users who generate information that catches the interest of fellow users/readers) and, then, adopt a Random Walk technique to find starters. The authors of [47] analyze the main properties of the nodes within a single OSN that connect the peripheral nodes and the peripheral groups with the rest of the network. The authors call these nodes bridging nodes or, simply, bridges. Clearly, here, the term "bridge" is used with a meaning totally different from that adopted in our paper. The authors base their analysis on the study of the theoretical properties of their model. In [25], the authors propose a predictive model that maps social media data to tie strength. This model is built on a dataset of social media ties and is capable of distinguishing between strong and weak ties with a high accuracy. Moreover, the authors illustrate how tie strength modeling can improve social media design elements, such as privacy controls, message routing, friend introductions and information prioritization. The authors of [60] present a model for predicting the closeness of professional and personal relationships of OSN users on the basis of their behavior in the OSNs joined by them. In particular, they analyze how the behavior of users on an OSN reflects the strength of their relationships with other users w.r.t. several factors, such as profile commenting and mutual connections.

A preliminary study about SISs and bridges has been done in [11]. However, it has been carried out by investigating samples extracted through classical crawling techniques. Berlingerio et al. [5, 6], Dai et al. [17], Mucha et al. [46], and Kazienko et al. [30] present approaches in the field of multidimensional networks. These networks can be seen as a specific case of a SIS in which each social network is specific for one kind of relationship and social networks strongly overlap. Multidimensional social networks are known as multislice networks in the literature [46].

Concerning the originality of our paper w.r.t. the above literature, we note that none of the above studies analyzes the main features of SIS. By contrast, in our paper, we provide a deep analysis on bridges, which are the key concept of a SIS.

3 The Bridge Driven Search Crawler

As pointed out in the introduction, the first main purpose of this paper is to investigate crawling strategies for a SIS. These must be able to extract not only connections among the accounts of different users in the same social network but also interconnections among the accounts of the same user in different social networks. Several crawling strategies for single social networks have been proposed in the literature. Among these strategies, two very popular ones are BFS [62] and RW [41]. The former implements the classical Breadth First Search visit, the latter selects the next node to be visited uniformly at random among the neighbors of the current node. A more recent strategy is MH [27]. At each iteration it randomly selects a node w from the neighbors of the current node v. Then, it randomly generates a number p belonging to the real interval $[0, 1]$. If $p \leq \frac{\Gamma(v)}{\Gamma(w)}$, where $\Gamma(v)$ ($\Gamma(w)$, resp.) is the outdegree of v (w, resp.), then it moves from v to w. Otherwise, it stays in v. The pseudocode of this algorithm is shown in Algorithm 1. Observe that the higher the degree of a node, the higher the probability that MH discards it.

In the past, these crawling strategies were deeply investigated when applied on a single social network. This analysis showed that none of them is always better than the others. Indeed, each of them can be the optimal one for a specific set of analyses. However, no investigation about the application of these strategies in a SIS has been carried out. Thus, we have no evidence that they are still valid in this new context. To reason about this, let us start by considering a structural peculiarity of a SIS, i.e. the existence of *bridges*, which, we recall, are those nodes of the graph corresponding to users who joined more than one social network and explicitly declared their different accounts. We expect that these nodes play a crucial role in the crawling of a SIS as they allow the crossing of different social networks, discovering the SIS intrinsic nature (related to interconnections). Bridges are not "standard" nodes, due to their role; thus, we cannot see a SIS just as a huge social network. Besides these intuitive considerations about bridges, we can help our reasoning also with two results obtained in [11], for which: *Fact (i)* the fraction of bridges in a social network

Algorithm 1: MH

Notation We denote by $\Gamma(x)$ the outdegree of the node x
Input s: a seed node
Output *SeenNodes*, *VisitedNodes*: a set of nodes
Constant n_{it} {The number of iterations}
Variable v, w: a node
Variable p: a number in the real interval (0,1)

SeenNodes:=∅, *VisitedNodes*:=∅
insert s into *SeenNodes* and *VisitedNodes*
insert all the nodes adjacent to s into *SeenNodes*
$v = s$
for $i := 1$ to n_{it} **do**
  let w be one of the nodes adjacent to v selected uniformly at random
  generate uniformly at random a number p in the real interval (0,1)
  if $(p \leq \frac{\Gamma(v)}{\Gamma(w)})$ **then**
    $v = w$
    insert w into *VisitedNodes*
    insert all the nodes adjacent to w into *SeenNodes*
  end if
end for

is low, and *Fact (ii)* bridges have high degrees on average. This is confirmed by the experimental results presented in Table 8.

Now, the question is: What about the capability of existing crawling strategies of finding bridges? The deep knowledge about BFS, RW and MH, provided by the literature, allows us to draw the following conjectures:

- BFS tends to explore a local neighborhood of the seed it starts from. As a consequence, if bridges are not present in this neighborhood or their number is low (and this is highly probable due to Fact *(i)*), the crawled sample fails in covering many social networks. Furthermore, it is well known that BFS tends to favor power users and, therefore, presents bias in some network parameters (e.g., the average degree of the nodes of the crawled portions are overestimated [36]).
- Differently from BFS, RW does not consider only a local neighborhood of the seed. In fact, it selects the next node to be visited uniformly at random among the neighbors of the current node. Again, due to Fact *(i)*, the probability that RW selects a bridge as the next node is low. As a consequence, the crawled sample does not cover many social networks and, if more than one social network is represented in it, the coupling degree of the crawled portions of social networks is low. Finally, analogously to BFS, RW tends to favor power users and, consequently, to present bias in some network parameters [36]. This feature only marginally influences the capability of RW to find bridges because, in any case, their number is very low.
- MH has been conceived to unfavor power users and, more in general, nodes having high degrees, which are, instead, favored by BFS and RW. It performs very well in a single social network [27] especially in the estimation of the average degree of nodes. However, due to Fact *(ii)*, it will penalize bridges. As a consequence, the sample crawled by MH does not cover many social networks present in the SIS.

In sum, from the above reasoning, we expect that both BFS, RW and MH are substantially inadequate in the context of SISs. As it will be described in Sect. 3.2, this conclusion is fully confirmed by a deep experimental campaign, which clearly highlights the above drawbacks. Thus, we need to design a specific crawling strategy for SISs. This is a matter of the next section.

3.1 BDS Crawling Strategy

In the design of our new crawling strategy, we start from the analysis of some aspects limiting BFS, RW, and MH in a SIS, to overcome them. Recall that BFS performs a Breadth First Search on a local neighborhood of a seed. Now, the average distance between two nodes of a single social network is generally less than the one between two nodes of different social networks. Indeed, to pass from a social network to another, it is necessary to cross a bridge, and as bridges are few, it may be necessary to generate a long path before reaching one of them. As a consequence, the local neighborhood considered by BFS includes one or a small number of social networks. To overcome this problem, a Depth First Search, instead of a Breadth First Search, can be done. For this purpose, the way of proceeding of RW and MH may be included in our crawling strategy. However, because the number of bridges in a social network is low, the simple choice to go in-depth blindly does not favor the crossing from a social network to another. Even worse, because MH penalizes the nodes with a high degree, it tends to unfavor bridges, rather than to favor them. Again, in the above reasoning, we have exploited Facts *(i)* and *(ii)* introduced in the previous section.

A solution that overcomes the above problems consists in implementing a "non-blind" depth first search in such a way as to favor bridges in the choice of the next node to visit. This is the choice we do, and the name we give to our strategy, i.e., *Bridge-Driven Search* (BDS, for short), clearly reflects this approach. However, in this way, it becomes impossible to explore (at least partially) the neighborhood of the current node because the visit proceeds in-depth very quickly and, furthermore, as soon as a bridge is encountered, there is a cross to another social network. The overall result of this way of proceeding is an extremely fragmented crawled sample. To address this problem, given the current node, our crawling strategy explores a fraction of its neighbors before performing an in-depth search of the next node to visit.

To formalize our crawling strategy, we need to introduce the following parameters:

- *nf* (*node fraction*). It represents the fraction of the *non-bridge* neighbors of the current node that should be visited. It ranges in the real interval (0,1]. For example, when *nf* is equal to 1, our strategy selects all the neighbors of the current node except the bridge ones. This parameter is used to tune the portion of the current node neighborhood that has to be taken into account and, hence, it balances the breadth and depth of the visit.

Algorithm 2: BDS

```
Notation We denote by N(x) the function returning the number of the non-bridge neighbors of the node x, and
    by B(x) the function returning the number of the bridges belonging to the neighborhood of x
Input s: the seed node
Output SeenNodes, VisitedNodes: a set of nodes
Constant n_it {The number of iterations}
Constant nf {The node fraction parameter}
Constant bf {The bridge fraction parameter}
Constant btf {The bridge tuning factor parameter}
Variable v, w: a node
Variable p: a number in the real interval (0,1)
Variable c: an integer number
Variable NodeQueue: a queue of nodes
Variable BridgeSet: a set of bridge nodes
SeenNodes:=∅, VisitedNodes:=∅, NodeQueue:=∅, BridgeSet:=∅
insert s into NodeQueue
for i := 1 to n_it do
    poll NodeQueue and extract a node v
    insert v into VisitedNodes
    insert all the nodes adjacent to v into SeenNodes
    if (B(v) ≥ 1) then
        clear NodeQueue
        c = 0
        while (c < ⌈bf · B(v)⌉) do
            let w be one of the bridge nodes adjacent to v not in BridgeSet selected uniformly at random
            generate uniformly at random a number p in the real interval (0,1)
            if (p · btf ≤ N(v)/N(w)) then
                insert w into NodeQueue and BridgeSet
                c = c + 1
            end if
        end while
    else
        c = 0
        while (c < ⌈nf · N(v)⌉) do
            let w be one of the nodes adjacent to v selected uniformly at random
            generate uniformly at random a number p in the real interval (0,1)
            if (p ≤ N(v)/N(w)) then
                insert w into NodeQueue
                c = c + 1
            end if
        end while
    end if
end for
```

- *bf* (*bridge fraction*). It represents the fraction of the bridge neighbors of the current node that should be visited. Like *nf*, it ranges in the real interval (0,1]. Clearly, this parameter is greater than 0 to allow the visit of at least one bridge (if any), resulting in crossing to another social network.
- *btf* (*bridge tuning factor*). It is a real number belonging to [0,1] that allows the filtering of the bridges to be visited among the available ones, on the basis of their degree. Its role will be better explained in the following.

For instance, in a configuration with $nf = 0.10$ and $bf = 0.25$, our strategy visits 10 % of the non-bridge neighbors of the current node and 25 % of the bridge neighbors of the current node.

We are now able to formalize our crawling strategy. Its pseudo-code is shown in Algorithm 2.

The algorithm exploits two data structures: a queue *NodeQueue* of nodes and a set *BridgeSet* of bridges. The former contains the nodes detected during the crawling task and that should be visited later; the latter contains the bridges that have been already met during the visit. BDS starts its visit from a seed node s that is added into *NodeQueue*. At each iteration, a new node v is extracted from *NodeQueue*; v is inserted into *VisitedNodes*, whereas all the nodes adjacent to v are put into *SeenNodes* (Lines 4–6). After this, if v has at least one bridge as neighbor, then the visit proceeds towards one or more of the bridge neighbors of v, thus switching the current social network. For this reason, *NodeQueue* is cleared (Line 8). This is necessary because, if the next nodes are polled from *NodeQueue*, the visit is brought back to the old social network improperly.

Now, in Line 10, the algorithm computes how many bridges must be selected on the basis of its setting. After this, each of these bridges, say w, is selected uniformly at random among those not previously met in the visit; w is added into *NodeQueue* and into *BridgeSet* if and only if the ratio between the v's outdegree and w's outdegree is greater than or equal to $p \cdot btf$, where p is a real random number in [0,1] (Line 13). Observe that this condition is similar to that adopted by MH to drive the selection of nodes on the basis of their degrees. In particular, when $btf = 1$, this condition coincides with the one of MH, disadvantaging high-degree bridges; when $btf = 0$, no filtering on the bridge degree is done. Clearly, values ranging from 0 to 1 result in an intermediate behavior. If no bridge has been discovered, then $\lceil nf \cdot N(v) \rceil$ non-bridges adjacent to v are randomly selected (Lines 20 and 21). Such nodes are selected according to the policy of MH. They are added into *NodeQueue*. The algorithm terminates after n_{it} iterations.

As for Lines 20–27, it is worth pointing out that, differently from MH, which selects only one neighbor for each node (thus, performing an in-depth visit), BDS has also a component (i.e., *nf*), which allows it to select more than just one neighbor in such a way as to make it able to take the neighborhood of the current node into account. In this way, it solves one of the problems of RW and MH discussed above.

3.2 Experiments

In this section, we present our experiment campaign conceived to determine the performances of BDS and to compare it with BFS, RW and MH when they operate in a SIS. As we wanted to analyze the behavior of these strategies on a SIS, we had to extract not only connections among the accounts of different users in the same social network but also connections among the accounts of the same user in different social networks. To encode these connections, two standards encoding human relationships are generally exploited. The former is XFN (XHTML Friends Network) [61]. XFN simply uses an attribute, called `rel`, to specify the kind of relationship between two users. Possible values of `rel` are `me`, `friend`, `contact`, `co-worker`, `parent`, and so on. A (presumably) more complex alternative to XFN is FOAF (Friend-Of-A-Friend) [8]. A FOAF profile is essentially an XML file describing people, their links to other people and their links to created objects. The technicalities concerning these two standards have not to be handled manually by the user. As a matter of fact, each social network has suitable mechanisms to automatically manage them in a way transparent to users, who have simply to specify their relationships in a friendly fashion.

In our experiments, we consider a SIS consisting of four social networks, namely Twitter, LiveJournal, YouTube and Flickr. They are compliant with the XFN and FOAF standards and have been largely analyzed in Social Network Analysis in the past [15, 34, 44, 62]. We argue that the relatively small number of involved social networks, as a first investigation, is adequate, expecting that the more this number, the higher the gap between standard and specific crawling strategies.

For our experiments, we exploited a server equipped with a 2 Quad-Core E5440 processor and 16 GB of RAM with the CentOS 6.0 Server operating system. Collected data can be found at the URL http://www.ursino.unirc.it/ebsnam.html. (The password to open the archive is "84593453".)

3.2.1 Metrics

A first needed step was to define reasonable metrics able to evaluate the performances of crawlers operating on a SIS. Even though this point may appear very critical and prone to unfair choices, it is immediate to realize that the following chosen metrics are a good way to highlight the desired features of a crawling strategy operating in a SIS:

1. *Bridge Ratio (BR)*: this is a real number in the interval [0,1] defined as the ratio of the number of the bridges discovered to the number of all the nodes in the sample.
2. *Crossings (CR)*: this is a non-negative integer and measures how many times the crawler switches from one social network to another.
3. *Covering (CV)*: this is a positive integer and measures how many different social networks are visited by the crawler.
4. *Unbalancing (UB)*: this is a non-negative real number and is defined as the standard deviation of the percentages of nodes discovered for each social network w.r.t. the overall number of nodes discovered in the sample. Observe that Unbalancing ranges from 0, corresponding to the case in which each social network is sampled with an equal number of nodes, to a maximum value (for instance, 50 in case of 4 social networks), corresponding to the case in which all sampled nodes belong to a social network. For example, in a SIS consisting of four social networks, if the overall discovered nodes are 100 and the number of nodes belonging to each of the four social networks is 40, 11, 30, and 19, resp., then *UB* is equal to 12.68.
5. *Degree Bias (DB)*: this is a real number computed as the root mean squared error, for each social network of the SIS, of the average node degree estimated by the crawler and that estimated by MH, which is considered the best one in

estimating the node degree for a social network in the literature [27, 36]. If the crawled sample does not cover one or more social networks, then these are not considered in the computation of the Degree Bias.

As for the first three metrics, the higher their value, the higher the performance of the crawling strategy. By contrast, as for the fourth and the fifth metric, the lower their values, the higher the performance of the crawling strategy. Observe that Covering is related to the crawler capability of covering many social networks. Unbalancing measures the crawler capability of uniformly sampling all the social networks. Furthermore, observe that, even though one may intuitively think that a fair sampling should sample different social networks proportionally to their respective overall size, a similar behavior of the crawler results in incomplete samples in case of high variance of these sizes. Indeed, it may happen that small social networks are not represented in the sample or represented in an insufficient way. Bridge Ratio and Crossing are related to the coupling degree, while Degree Bias to the average degree. Finally, we note that the defined metrics are not completely independent from each other. For instance, if $BR = 0$, then CR and CV are also 0. Analogously, the value of CR influences both CV and UB.

Besides the evaluation of the crawling strategies on each of the above metrics, separately considered, it is certainly important to define a synthetic measure capable of capturing a sort of "overall" behavior of the strategies, possibly modulating the importance of each metric. A reasonable way to do this is to compute a linear combination of the five metrics, in which the coefficients reflect the importance associated with them. We call *Average Crawling Quality* (*ACQ*) this measure and define it as:

$$ACQ = w_{BR} \cdot \frac{BR}{BR_{max}} + w_{CR} \cdot \frac{CR}{CR_{max}} + w_{CV} \cdot \frac{CV}{CV_{max}} + w_{UB} \cdot (1 - \frac{UB}{UB_{max}}) + w_{DB} \cdot (1 - \frac{DB}{DB_{max}})$$

where BR_{max} (CR_{max}, CV_{max}, UB_{max}, DB_{max}, resp.) are upper bounds of Bridge Ratio (Crossings, Covering, Unbalancing, Degree Bias, resp.) that, in a comparative experiment, can be set to the maximum value obtained by the compared techniques, whereas $w_{BR}, w_{CR}, w_{CV}, w_{UB}$, and w_{DB} are positive real numbers belonging to [0, 1] such that $w_{BR} + w_{CR} + w_{CV} + w_{UB} + w_{DB} = 1$. Below, we deal with the problem of setting the values of these parameters.

3.2.2 Analysis of BFS, RW and MH

In this section, we analyze the performances of BFS, RW and MH, when applied on a SIS. For this purpose, we randomly chose four seeds, each belonging to one of the social networks of our SIS, and, for each crawling strategy, we run the corresponding crawler one time for each of the four seeds. The number of iterations of each crawling run was 5,000. The overall numbers of seen nodes returned by MH,

Table 1 Performances of MH, BFS and RW

		Twitter	YouTube	Flickr	LiveJournal	Overall
MH	Bridge Ratio	0.0028	0.0038	0.0	0.0024	0.0025
	Crossing	5	3	0	4	3
	Covering	2	2	1	3	2
	Unbalancing	32.5503	40.6103	50	31.1253	38.5715
	Degree Bias					0
	Twitter Avg. Deg.	*37.9189*	*37.0789*	–	*38.2842*	*37.76067*
	YouTube Avg. Deg.	*9.6064*	*10.0379*	–	*10.4144*	*10.01957*
	Flickr Avg. Deg.	–	–	*104.3497*	–	*104.3497*
	LiveJournal Avg. Deg.	–	–	–	*40.5604*	*40.5604*
BFS	Bridge Ratio	0.0008	0.0054	0.0	0.0	0.0016
	Crossing	7	43	0	0	12.5
	Covering	2	3	1	1	1.75
	Unbalancing	49.9350	42.0286	50	50	47.9909
	Degree Bias					103.9339
	Twitter Avg. Deg.	*21.92*	–	–	–	*21.92*
	YouTube Avg. Deg.	–	*9.5106*	–	–	*9.5106*
	Flickr Avg. Deg.	–	–	*311.6118*	–	*311.6118*
	LiveJournal Avg. Deg.	–	–	–	*40.0136*	*40.0136*
RW	Bridge Ratio	0.0	0.00067	0.0	0.0	0.00017
	Crossing	0	4	0	0	1
	Covering	1	3	1	1	1.5
	Unbalancing	50	40.1363	50	50	47.5341
	Degree Bias					180.8260
	Twitter Avg. Deg.	*39.0*	*41.9489*	–	–	*40.4745*
	YouTube Avg. Deg.	–	*12.5081*	–	–	*12.5081*
	Flickr Avg. Deg.		*476.6446*	*455.3002*	–	*465.9724*
	LiveJournal Avg. Deg.	–	–	–	*43.3333*	*43.3333*

RW and BFS were 135,163, 941,303 and 726,743, respectively. The high variance of these numbers is not surprising because it is intrinsic in the way of proceeding of these algorithms.

In Table 1, we show the values of our metrics, along with the values of the other parameters we consider particularly significant (i.e., the average degree of the nodes of each social network), obtained for the four runs of MH, BFS and RW, respectively.

From the analysis of this table, we can draw the following conclusions:

- The value of *BR* is very low for all the crawling strategies. For MH there are on average 2.5 bridges for each 1,000 crawled nodes. BFS behaves worse than MH, and RW is the worst one. This behavior can be explained by the theoretical observations about BFS, MH and RW provided in Sect. 3.
- The value of *CR* is generally low for all the crawling strategies. RW shows again the worst value. This result is clearly related to the low value of *BR*, because the

few discovered bridges do not allow the crawlers to sufficiently cross different social networks.

- The value of *CV* is quite low for all the strategies. On average only two of the social networks of the SIS are visited. Also this result is related to the low values of *BR* and *CR*.
- Even though *BR*, *CR* and *CV* are generally low for all social networks, this trend is mitigated for YouTube when BFS and RW are adopted. This can be explained by the fact that the central concept in YouTube is *channel*, rather than profile. A channel has generally associated the links with the profiles of the corresponding owners present in the other social networks. As a consequence, YouTube tends to behave as a "hub" among the other social networks. This implies that the number of bridges in YouTube is higher than in the other social networks; in its turn, this implies an increase in *BR*, *CR* and *CV*. This trend is not observed for MH because this crawling technique tends to unfavor high-degree nodes, and often bridges have this characteristic (see *Fact (ii)* in Sect. 3).
- The value of *UB* is very high for all the strategies, very close to the maximum one (i.e., 50). This indicates that, as far as this metric is concerned, they behave very badly. Indeed, it happens that they often stay substantially bounded in the social network of the starting seed. This result can be explained (*i*) by the fact that *UB* is influenced by *CR*, which, in turn, is influenced by *BR*, and (*ii*) by the previous conclusions about *BR* and *CR*.
- As for the average degrees of nodes, it is well known that MH is the crawling strategy that best estimates them in a single social network [27, 36]. From the analysis of Table 1, we observe that when MH starts from a seed it generally stays for many iterations in the corresponding social network (this is witnessed by the high values of *UB*). As a consequence, we can assume that the average degrees are those of reference for the social networks of the SIS, provided that at least one run of MH starting from each social network is performed. This conclusion is further enforced by observing that (as shown in Table 1) MH is capable of estimating the average degree of nodes even for social networks different from that of the seed (whose number of nodes in the sample is quite low). Basing on these reference values, we detect that BFS presents a high value of *DB*. This is well known in the literature for a scenario consisting of a single social network [36], and we confirm this conclusion also in the context of SISs. The performance of RW is even worse than that of BFS.

In sum, we may conclude that the conjectures given above about the unsuitability of BFS, MH and RW to operate on a SIS are fully confirmed by our experiments. Now we have to see how our crawling strategy performs in this scenario. This is the matter of the next section.

Table 2 Performances of BDS for different values of *nf*

bf = 0.25, *btf* = 0.25	*nf* = 0.02	*nf* = 0.10	*nf* = 0.25	*nf* = 0.50
Bridge Ratio	0.0488	0.0965	0.1516	0.0814
Crossing	28	286	639	410
Covering	3	4	4	4
Unbalancing	26.1299	13.5103	12.2921	42.2080
Degree Bias	N.A.	19.0145	68.7413	123.705
Twitter Avg. Deg.	*40.5714*	*42.0383*	*41.0812*	*42.1161*
YouTube Avg. Deg.	*12.079*	*11.9278*	*12.8508*	*14.3556*
Flickr Avg. Deg.	*240.8333*	*129.5833*	*153.6764*	*331.5379*
LiveJournal Avg. Deg.	N.A.	*68.6234*	*168.8152*	*138.3333*

3.2.3 Analysis of BDS

To analyze BDS we performed a large set of experiments. In this section, we present the most significant ones to evaluate the impact of *nf*, *bf* and *btf*. In the configurations considered in these experiments we performed 5,000 iterations. The number of obtained seen nodes ranges from 15,585 to 473,122.

Impact of *nf*

We first evaluate the role of *nf* on the behavior of BDS. For this purpose we have fixed the other two parameters *bf* and *btf* to 0.25, and we have assigned to *nf* the following values: 0.02, 0.10, 0.25 and 0.50 (the reasons underlying the choice of discarding lower or higher values will be clear below). The results of this experiment are shown in Table 2. From the analysis of this table we observe that very low values of *nf* (i.e., *nf* about 0.02) lead to a significant decrease in *BR* and *CR*. Furthermore not all the social networks of the SIS are sampled. This behavior can be explained by the fact that, when *nf* is very low, BDS behaves as RW. This has a very negative influence on *UB*, because *BR* and *CR* influence *UB*, and does not allow the computation of *DB*, because not all social networks of the SIS are covered. As a consequence, we have decided not to report values of *nf* lower than 0.02.

By contrast, for high values of *nf* (i.e., *nf* about 0.50) we observe that *BR*, *CR* and *CV* show satisfying values. However, in this case, we obtain the worst values of *UB* and *DB*, registering for these metrics a behavior of BDS similar to that of BFS (as a matter of fact, *nf* = 0.50 implies that 50 % of the non-bridge neighbors of each node are visited). The high *UB* is explained by the fact that, even though the visit involved all the four social networks, this did not happen in a uniform fashion and some social networks have been sampled much more than the others. As for *DB*, in this case, BDS shows a behavior even worse than BFS because the presence of a high number of bridges in the sample causes the increase in the estimated average degree of the social networks (recall Fact *(ii)* introduced in Sect. 3). For this reason we do not report values of *nf* higher than 0.50.

Table 3 Performances of BDS for different values of *bf*

$nf = 0.10$, $btf = 0.25$	$bf = 0.25$	$bf = 0.50$	$bf = 0.75$	$bf = 1.00$
Bridge Ratio	0.0965	0.0875	0.0815	0.0824
Crossing	286	264	270	266
Covering	4	4	4	4
Unbalancing	13.5103	18.4622	8.4756	16.1804
Degree Bias	19.0145	21.8163	53.9505	50.8895
Twitter Avg. Deg.	*42.0383*	*40.76*	*40.4362*	*40.3348*
YouTube Avg. Deg.	*11.9278*	*12.2291*	*11.9679*	*11.7807*
Flickr Avg. Deg.	*129.5833*	*147.428*	*157.427*	*156.6471*
LiveJournal Avg. Deg.	*68.6234*	*46.4071*	*134.4459*	*127.82*

The reasoning above suggests that, to cover all the social networks of the SIS, *nf* should be higher than 0.02. However, to obtain acceptable *DB* values, it should be lower than 0.50. For this reason, we decided to fix *nf* to the intermediate value 0.10 in the study of *bf* and *btf*. In fact, this value shows a good tradeoff w.r.t. all considered metrics.

Impact of *bf*

To evaluate the impact of *bf* on the behavior of BDS we fixed *nf* to 0.10 and *btf* to 0.25. We assigned to *bf* the following values: 0.25, 0.50, 0.75 and 1. We set 0.25 as the lower bound for *bf* because, by a direct analysis on the sample obtained by setting $bf = 1$, we saw that the maximum number of bridges adjacent to a node was 4. The values of the metrics obtained in this case are reported in Table 3.

From the analysis of this table we can observe that, as for the first four metrics, the obtained results are satisfying and comparable for each value of *bf*. This is a further confirmation that fixing $nf = 0.10$ allows BDS to cover all the social networks of the SIS in a satisfactory way. The only discriminant for *bf* seems to be *DB* because the increase in *bf* leads to an increase in the average node degree. This trend can be explained as follows. When a user has more adjacent bridges (less than 4, in our sample), for $bf = 0.25$, BDS selects only one of them. In this case, with the highest probability, the selected bridge will be the one with the lowest degree (see Line 13 in Algorithm 2). In the same case, if $bf = 1$, then BDS selects all adjacent bridges and, therefore, also those having the highest degrees. From the assortativity property [48], it is well known that high-degree users are often connected with other high-degree users. All these facts imply that the average degree of the sampled nodes increases. This explains the seemingly "strange" decrease in *BR* observed when *bf* increases. In fact, because the number of adjacent bridges is very limited, when the number of adjacent nodes increases, the fraction of bridges present in it (i.e., *BR*) decreases. All these reasonings suggest that an increase in *bf* causes an increase in *DB* and a decrease in *BR*. All the other metrics do not show significant variations. For this reason, in these experimental campaigns, we fixed *bf* to 0.25.

Table 4 Performances of BDS for different values of *btf*

$nf = 0.10$, $bf = 0.25$,	$btf = 0.00$	$btf = 0.25$	$btf = 0.50$	$btf = 0.75$	$btf = 1.00$
Bridge Ratio	0.0771	0.0965	0.0715	0.07612	0.0705
Crossing	245	286	226	258	183
Covering	4	4	4	4	4
Unbalancing	20.1134	13.5103	9.9273	11.8239	14.8959
Degree Bias	140.4817	19.0145	32.8467	48.7946	29.8698
Twitter Avg. Deg.	*40.2287*	*42.0383*	*39.7203*	*39.0245*	*39.6141*
YouTube Avg. Deg.	*11.8225*	*11.9278*	*11.1113*	*11.5115*	*11.3697*
Flickr Avg. Deg.	*379.377*	*129.5833*	*118.1566*	*124.1844*	*133.0608*
LiveJournal Avg. Deg.	*97.9286*	*68.6234*	*104.7473*	*136.0927*	*92.8981*

Impact of *btf*

In this experimental campaign, we fixed *nf* to 0.10 and *bf* to 0.25. We considered the following values for *btf*: 0, 0.25, 0.50, 0.75, and 1. *btf* can be seen as a filter on the bridge degrees. In particular, if $btf = 0$, then there is no constraint on the degrees of the bridges to select. If $btf = 1$, then BDS behaves as MH and, therefore, favors the selection of those bridges whose degree is lower than or equal to that of the current node. The other bridges are selected with a probability that decreases with the increase in their degree. The values of the metrics measured in this experimental campaign are reported in Table 4.

From the analysis of this table, it is evident that, when $btf = 0$, *DB* is high. This can be explained by the fact that, in this case, all bridges (even those with very high degree) may be equally selected. For the other values of *btf*, the overall performances of BDS do not present significant differences because all these values allow high-degree bridges to be filtered out. From a direct analysis on our samples we verified that the average degree of bridges is at most four times that of non-bridges. Setting $btf = 0.25$ allows that, even in the worst case (i.e., $p = 1$ in Line 13 of Algorithm 2), the bridges having a degree lower than or equal to the average degree of non-bridges are generally selected. This way, high-degree bridges are unfavored whereas the others are highly favored. For this reason, in these experimental campaigns, we fixed *btf* to 0.25.

3.2.4 Average Crawling Quality

So far we have analyzed the behavior of BDS w.r.t. the five metrics separately considered. To compare our strategy with the other three ones, it is more important to study their "overall" behavior by using the metric *ACQ*, which aggregates all the five metrics considered previously. Here, we have to deal with the problem of setting the coefficients of the linear combination, namely $w_{BR}, w_{CR}, w_{CV}, w_{UB}$, and w_{DB}, present in the definition of *ACQ*.

Table 5 *ACQ* for the different parameter configurations of BDS

Configuration	*ACQ* (same weights)	*ACQ* (different weights)
nf = 0.02 bf = 0.25 btf = 0.25	0.42	0.36
nf = 0.10 bf = 0.25 btf = 0.00	0.48	0.42
nf = 0.10 bf = 0.25 btf = 0.25	0.84	0.86
nf = 0.10 bf = 0.25 btf = 0.50	0.67	0.57
nf = 0.10 bf = 0.25 btf = 0.75	0.66	0.57
nf = 0.10 bf = 0.25 btf = 1.00	0.64	0.55
nf = 0.10 bf = 0.50 btf = 0.25	0.68	0.63
nf = 0.10 bf = 0.75 btf = 0.25	0.68	0.59
nf = 0.10 bf = 1.00 btf = 0.25	0.64	0.57
nf = 0.25 bf = 0.25 btf = 0.25	0.73	0.67
nf = 0.50 bf = 0.25 btf = 0.25	0.46	0.44

We start by assigning the same weight to all metrics, i.e., we set $w_{BR} = w_{CR} = w_{CV} = w_{UB} = w_{DB} = 0.2$. Then, we measure the value of *ACQ* for all the configurations of the parameters of BDS examined in Tables 2, 3 and 4. Obtained values are reported in the second column of Table 5. From the analysis of this column, we can verify that the configuration of BDS that guarantees the best tradeoff among the various metrics is $nf = 0.10$, $bf = 0.25$, $btf = 0.25$.

Observe that, at the beginning of Sect. 3.2 we showed that the defined metrics are not completely independent of each other. In fact, *BR* influences *CR* and *CV*, whereas *CR* influences *CV* and *UB*. As a consequence, it is reasonable to associate different weights with the various metrics by assigning the higher values to the most influential ones. To determine these values, we use an algorithm that takes inspiration from the Kahn's approach for topological sorting of graphs [29]. In particular, we first construct the *metric Dependency Graph*. It has a node n_{M_i} for each metric M_i. There is an edge from n_{M_i} to n_{M_j} if the metric M_i influences the metric M_j. A weight is associated with each node. Initially, we set all the weights to 0.20 (Fig. 1). We start from a node having no outgoing edges and split its weight (in equal parts) among itself and the nodes it depends on.[1] Then, we remove all the incoming edges. We repeat the previous tasks until all the nodes of the graph have been processed. By applying this approach we obtain the following configuration of weights: $w_{BR} = 0.45$, $w_{CR} = 0.18$, $w_{CV} = 0.07$, $w_{UB} = 0.10$, $w_{DB} = 0.20$. Observe that the node processing order is not unique because more than one node with no outgoing edge exists. However, it is easy to verify that the final metric weights returned by our algorithm do not depend on the adopted node processing order.

[1]Clearly, if a node has no incoming edges, it maintains its weight.

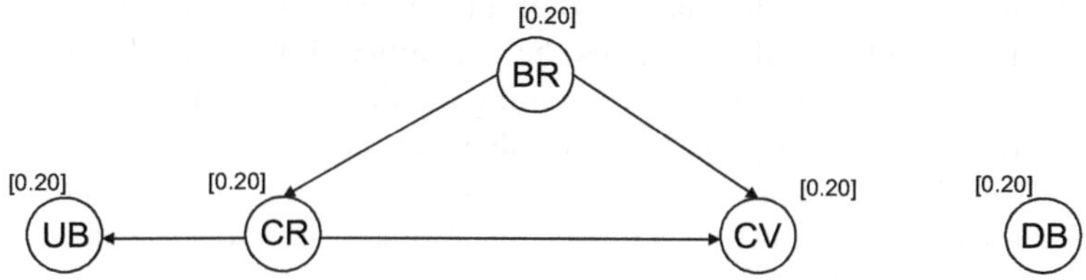

Fig. 1 The Dependency Graph concerning our metrics

Table 6 Values of *ACQ* for the different crawling techniques

Technique	*ACQ* (same weights)	*ACQ* (different weights)
MH	0.34	0.26
BFS	0.18	0.12
RW	0.08	0.03
BDS	0.84	0.86

Now, we measure *ACQ* with this new weight setting. The obtained results are reported in the third column of Table 5. Also in this experiment, the setting $nf = 0.10$, $bf = 0.25$, $btf = 0.25$ of the parameters of BDS shows the best performance.

However, with regard to this result, there are applications in which some metrics are more important than the others. BDS is highly flexible and, in these cases, allows the choice of the configuration that favors those metrics. For instance, if a user performs link mining in a SIS, the most important metric is *CR* because it is an index of the number of links between different social networks present in the crawled sample. By contrast, *DB* does not appear particularly relevant. In this case, the configuration $nf = 0.25$, $bf = 0.25$, $btf = 0.25$, is chosen because it guarantees the maximum *CR* even though the corresponding Bias Degree is quite high (see Table 2). As a second example, if it is desired a crawled sample in which all the social networks of the SIS are represented in the most uniform way, it is suitable to adopt the configuration $nf = 0.10$, $bf = 0.75$, $btf = 0.25$ that guarantees the best *UB* (see Table 3).

We now compare BDS, BFS, RW, and MH when they operate on a SIS. In this comparison, as for BFS, RW and MH we selected the overall values (see the last column of Table 1). As for BDS we adopted the configuration $nf = 0.10$, $bf = 0.25$, $btf = 0.25$. The results of this comparison, obtained by computing *ACQ* with the two weight settings, are reported in Table 6.

Interestingly enough, even the lowest value of *ACQ* obtained for BDS (obtained with the configuration $nf = 0.02, bf = 0.25, btf = 0.25$) is higher than that obtained for MH and much higher than those obtained for BFS and RW. This shows that BDS guarantees always the best performance.

From the analysis of these values and of those reported in Tables 1, 2, 3, and 4, it clearly emerges that, when operating on a SIS, BDS highly outperforms the other approaches. The only exception is MH for *DB* because, according to [27,36], we have assumed that MH is the best method to estimate the average node degree. However, also for this metric, BDS obtains very satisfactory results. As a final remark we highlight that, besides the capability shown by BDS of crossing through different social networks, overcoming the drawbacks of compared crawler strategies, BDS presents a good behavior also from an *intra-social-network* point of view. This claim is supported from both the results obtained for *DB*, and the consideration that our crawling strategy, in absence of bridges, can be located between BFS and MH, producing intra-social-network results that reasonably cannot differ significantly from the above strategies.

4 Experiences

As pointed out in the introduction, the second main purpose of this paper is to exploit BDS for investigating the main features of bridges and SISs. This section is devoted to this analysis and is organized in such a way that each subsection investigates a specific aspect of SISs, namely the degree of bridges and non-bridges, the relationships between bridges and power users, the possible existence of a bridge backbone and the analysis of bridge centrality.

To perform the analyses of this section, we collected ten samples using BDS. We performed each investigation described below on each sample and, then, we averaged the obtained values on all of them. Therefore, each measure reported below is the average of the values obtained on each sample.

4.1 Distributions of Bridge and Non-bridge Degrees

In this section, we analyze the distributions of node degrees. For this purpose, we compute the Cumulative Distribution Function (CDF) of the degree of bridges and non-bridges. This function describes the probability that the degree of a node is less than or equal to a given value x. The CDFs for bridges and non-bridges are shown in Fig. 2.

By analyzing this figure, we can see that, fixed a degree d, the probability that a bridge has more than d contacts is higher than that of a non-bridge, for any d. As a consequence, we can state that a bridge has more contacts than a non-bridge, in average.

Again, observing the CDF trend for both bridges and non-bridges, it seems that the corresponding degrees follow a power law distribution. To verify this conjecture, in Fig. 3 we plot the Probability Distribution Function (PDF) of the degree of bridges and non-bridges. A visual analysis of the PDF trend already confirms our conjecture. To refine our analysis, we compute the best power law fit using the maximum likelihood method [16]. Table 7 shows the estimated power law coefficients, along with the Kolmogorov-Smirnov goodness-of-fit metrics, for the distributions into consideration. In particular, α is the exponent of the theoretical power law function

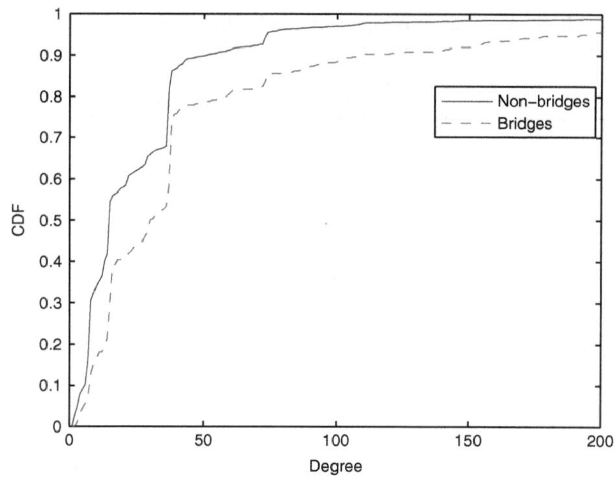

Fig. 2 Cumulative Distribution Function for bridges and non-bridges

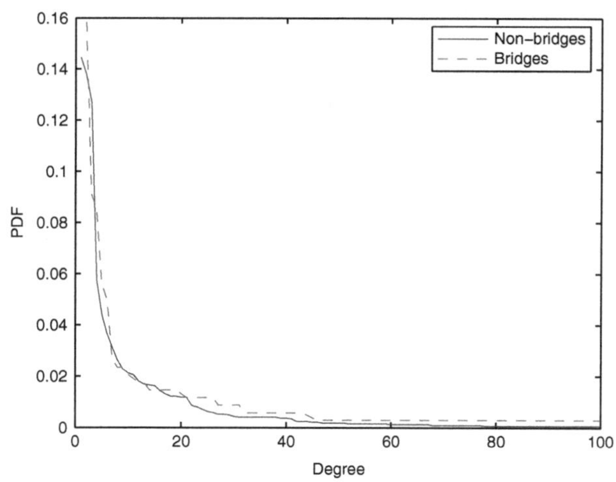

Fig. 3 Probability Distribution Function for bridges and non-bridges

Table 7 Power law coefficient estimation for the PDF of bridges and non-bridges

	α	D
Non-bridges	2.18	0.09
Bridges	2.31	0.15

that best approximates the real one, whereas D is the maximum distance between the theoretical function and the real one. The shown results, and in particular the low value of the Kolmogorov-Smirnov goodness-of-fit metric, confirm that the degrees of bridges and non-bridges follow a power law distribution.

To deepen our analysis, we compute the average degrees of bridges and non-bridges, along with the corresponding standard deviations, for each social network of the SIS. The obtained results are presented in Table 8. From the analysis of this table we can observe that: (1) the standard deviations of bridges and non-bridges are generally high; this can be explained by the power law distribution of PDF; (2) both

Table 8 Analysis of bridge and non-bridges degrees for the whole SIS and its social networks

	AVG		STD	
	Bridges	Non-bridges	Bridges	Non-bridges
YouTube	15.13	11.67	9.00	8.30
Flickr	315.47	97.01	834.41	119.96
LiveJournal	159.93	49.21	145.75	59.66
Twitter	44.21	41.75	20.84	19.73
All	66.69	26.82	282.10	41.13

the average degree and the degree standard deviation of bridges are higher than the corresponding ones of non-bridges for all the social networks of the SIS; in other words, this trend, valid for the SIS in the whole, is general and not specific for some social network.

In other words, BDS confirms the same results obtained in [11] with the other crawling techniques, and, in turn, this represents a further confirmation of Fact *(ii)* introduced in Sect. 3.

Continuing the analysis of Fig. 2, we can observe a relevant discontinuity of the CDF for both bridges and non-bridges around the degrees 35–40. Indeed, we have that the probability to find a bridge with less than 35 contacts is about 0.5, and that this probability becomes 0.75 for bridges with less than 40 contacts. The same trend occurs for non-bridges. This may be explained by considering that there exist two typologies of social network users. The former is composed by users who joined a social network for a short time, adding a limited (less than 40) number of friends. The latter refers to users who are active and, therefore, have an increasing number of contacts. The former typology raises the CDF values in the initial range (say 0–40), generating the observed discontinuity.

4.2 Bridges and Power Users

As seen in the previous section, bridges have an average degree higher than that of non-bridges. A question arises spontaneously: Are bridges power users? As a matter of fact, according to the definition given in [56], power users are nodes having a degree higher than the average degree of the other nodes. Indeed, if bridges were power users, for their detection it is possible to exploit the techniques for power user extraction already proposed in the literature. In the previous experiment, we measured that average degree of bridges is 66.69, whereas that of non-bridges is 26.82. The average degree of all nodes is 30.67, which is the reference value for classifying a node as power users. Looking at the average degrees, we may expect that bridges are actually power users. However, because degrees follow a power law distribution, this conjecture may be wrong.

To solve this question, we have to understand how much the set of power users and that of bridges overlap. Specifically, we denote by P the set of power users and

Fig. 4 Overlapping between bridge set and power user set

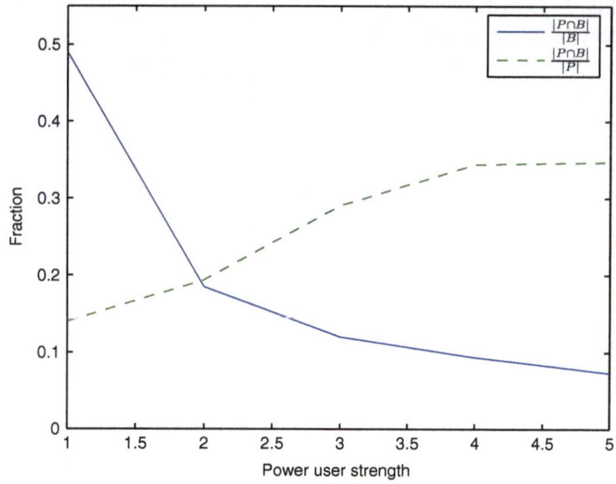

by B the set of bridges. Then, we measure the fraction of bridges that are power users and the fraction of power users that are bridges. The obtained results are: $\frac{|P \cap B|}{|B|} = 0.49$ and $\frac{|P \cap B|}{|P|} = 0.14$. They show that half of bridges are power users, whereas only few power users are bridges.

To better understand this phenomenon, we extend the concept of power user by introducing the notion of the strength of a power user. In particular, we say that a power user is an *s-strength power user* if its degree is s times higher than the average degree of nodes. Clearly, a standard power user corresponds to a 1-strength power user. Now, we compute $\frac{|P \cap B|}{|B|}$ and $\frac{|P \cap B|}{|P|}$ for increasing values of the strength of power user. The results of this experiment are shown in Fig. 4.

Here, it is possible to see that initially half of the bridges are power users. However, the percentage of bridges that are s-strength power users decreases as s increases. This allows us to conclude that bridges are not "strong" power users. Viceversa, the percentage of power users that are bridges increases as s increases. This allows us to conclude that the probability of finding a bridge among the strongest power users is higher than that of finding a bridge among the weak power users. However, this probability is never higher than 0.35.

In any case, both the (decreasing) trend of $\frac{|P \cap B|}{|B|}$ and the low values of $\frac{|P \cap B|}{|P|}$ allow us to conclude that there does not exist a meaningful correlation between bridges and power users.

4.3 Ties Among Bridges and Non-bridges

In this analysis, we aim at studying whether bridges have preferential ties among them, i.e., whether they are more likely to be connected to each other than to non-bridges. A possible way to carry out this verification is to compute the distribution of

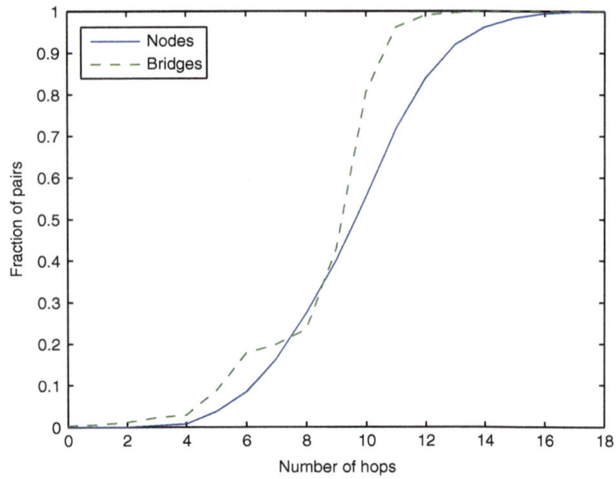

Fig. 5 Distribution of the lengths of the shortest paths among bridges and among nodes

the lengths of the shortest paths among bridges and among nodes in the SIS. Indeed, if these distributions are similar and the maximum lengths of the shortest paths connecting two nodes and two bridges are comparable, it is possible to conclude that no preferential tie exists among bridges. By contrast, if the maximum length of the shortest paths connecting two bridges is less than that of the shortest paths connecting two nodes and/or the distributions are dissimilar in the sense that the one of bridges raises much faster than the one of non-bridges, it is possible to conclude that bridges are likely to be connected to each other. The results of this experiment are shown in Fig. 5.

From the analysis of this figure, we can see that the distribution of bridge distance follows the same trend of that of node distance. Moreover, the effective diameter (90-th percentile of the distribution of the lengths of the shortest paths) measured for nodes and bridges is about 12 and 11, respectively. This allows us to conclude that no preferential connection favoring the link among bridges exists.

Interestingly enough, this experiment supplies us a further hint about the node to be used as seed in a crawling task for a SIS: Indeed, we cannot rely on a particular seed to enhance the percentage of bridges in a crawled sample, because a backbone among bridges does not exist.

4.4 Bridge Centrality

This experiment is devoted to analyze the centrality of bridges in a SIS. Centrality is one of the most important measures adopted in Social Network Analysis to investigate the features of nodes in a social network. Basically, there are four main centrality metrics, namely *degree*, *betweennes*, *closeness* and *eigenvector* [20]. In this experiment, we focus on *betweenness* because it computes the centrality of a

Table 9 Centrality of bridges and non-bridges

	Bridges	Non-bridges
Twitter	2,404	2,758
Flickr	643	652
LiveJournal	33	36
YouTube	3,779	5,685
All	23,100	12,156

node by quantifying how much it is important in guaranteeing the communication among other nodes and, therefore, how much it acts as bridge along the shortest paths of other nodes.

We expect that bridges have a high *betweenness* value in the whole SIS. In this experiment, we aim at verifying this intuition and, in the affirmative case, at studying if they maintain this property in the single social networks they joined.

For this purpose, we compute the *betweennes* of each node in our samples by means of *SNAP* (Stanford Network Analysis Platform) [56] and we average the corresponding values for bridges and non-bridges. The results are reported in Table 9.

By analyzing this table, we can observe that the intuition about the high *betweennes* of bridges in the whole SIS is fully confirmed. By contrast, in the single social networks, the values of *betweennes* of bridges are comparable or less than those of non-bridges. At a first glance, this result is unexpected because it appears immediate to think that a bridge can maintain its role of connector also in the single social networks joined by it. Actually, a more refined reasoning leads us to conclude that often bridges, just for their role, are at the borders of their social networks, and this partially undermines their capability to be central.

5 Conclusion

In this paper, first we have investigated the problem of crawling Social Internetworking Scenarios. We have started from the consideration that existing crawling strategies are not suitable for this purpose. In particular, we have analyzed the state-of-the-art techniques, which are BFS, RW and MH, showing experimentally that the above claim is true. On the basis of this result, by analyzing the reasons of the drawbacks of existing crawling strategies, we have designed a new one, called BDS (Bridge-Driven Search), specifically conceived for a SIS. We have conducted several experiments showing that, when operating in a SIS, BDS highly outperforms BFS, RW and MH, and arguing that BDS presents a good behavior also in intra-social-network crawling. Besides the overall conclusion mentioned above, we have seen that BDS is highly flexible as it allows a metric to be privileged over another one. After having validated BDS, we have exploited it to explore the emergent scenario of Social Internetworking from the perspective of Social Network Analysis. Being aware that the complete investigation of all the aspects of SISs is an extremely large task, we have identified the most basic structural peculiarity of these systems, i.e. bridges, and we have deeply studied it. We argue that most of the knowledge about the structural properties of SISs, and possibly about the behavioral aspects of users, starts from the adequate knowledge of bridges, which are the structural pillars of SISs.

We think that SIS analysis is a very promising research field and so we plan to perform further research efforts in the future. In particular, one of the most challenging issue is the improvement of the BDS crawling strategy in such a way that the values of *nf*, *bf* and *btf* dynamically change during the crawling activity to adapt themselves to the specificities of the crawled SIS. Moreover, we plan to investigate the possible connections of our approach with the information integration ones, as well as to deal with the privacy issue arising when crawling SISs. Another important future development regards the exploitation of BDS (or its evolutions) to perform a deeper investigation of SISs. In this context, it appears extremely promising to apply Data Warehousing, OLAP and Data Mining techniques on SIS samples derived by applying BDS to derive knowledge patterns about SISs.

Acknowledgements This work has been partially supported by the TENACE PRIN Project (n. 20103P34XC) funded by the Italian Ministry of Education, University and Research.

References

1. Agarwal N, Galan M, Liu H, Subramanya S (2010) WisColl: collective wisdom based blog clustering. Inf Sci 180(1):39–61
2. Ahn YY, Han S, Kwak H, Moon S, Jeong H (2007) Analysis of topological characteristics of huge online social networking services. In: Proceedings of the international conference on world wide web (WWW'07), Banff, Alberta. ACM, New York, pp 835–844
3. Backstrom L, Huttenlocher D, Kleinberg J, Lan X (2006) Group formation in large social networks: membership, growth, and evolution. In: Proceedings of the ACM SIGKDD international conference on knowledge discovery and data mining (KDD'06), Philadelphia. ACM, New York, pp 44–54
4. Berlingerio M, Coscia M, Giannotti F, Monreale A, Pedreschi D (2010) Towards discovery of eras in social networks. In: Proceedings of the workshops of the international conference on data engineering (ICDE 2010), Long Beach. IEEE, Los Alamitos, CA, USA, pp 278–281
5. Berlingerio M, Coscia M, Giannotti F, Monreale A, Pedreschi D (2011) Foundations of multidimensional network analysis. In: Proceedings of the international conference on advances in social networks analysis and mining (ASONAM 2011), Kaohsiung. IEEE, Los Alamitos, CA, USA, pp 485–489
6. Berlingerio M, Coscia M, Giannotti F, Monreale A, Pedreschi D (2011) The pursuit of hubbiness: analysis of hubs in large multidimensional networks. J Comput Sci 2(3):223–237
7. Bonneau J, Anderson J, Danezis G (2009) Prying data out of a social network. In: Proceedings of the international conference on advances in social network analysis and mining (ASONAM'09), Athens. IEEE, Los Alamitos, CA, USA, pp 249–254
8. Brickley D, Miller L (2012) The friend of a friend (FOAF) project. http://www.foaf-project.org/
9. Buccafurri F, Lax G, Nocera A, Ursino D (2012) Crawling social internetworking systems. In: Proceedings of the international conference on advances in social analysis and mining (ASONAM 2012), Istanbul. IEEE Computer Society, Los Alamitos, pp 505–509

10. Buccafurri F, Lax G, Nocera A, Ursino D (2012) Discovering links among social networks. In: Proceedings of the European conference on machine learning and principles and practice of knowledge discovery in databases (ECML PKDD 2012), Bristol. Lecture notes in computer science. Springer, Berlin, pp 467–482
11. Buccafurri F, Foti VD, Lax G, Nocera A, Ursino D (2013) Bridge analysis in a social internetworking scenario. Inf Sci 224:1–18
12. Carrington P, Scott J, Wasserman S (2005) Models and methods in social network analysis. Cambridge University Press, Cambridge
13. Catanese SA, De Meo P, Ferrara E, Fiumara G, Provetti A (2011) Crawling Facebook for social network analysis purposes. In: Proceedings of the international conference series on web intelligence, mining and semantics (WIMS'11), Sogndal. ACM, New York, pp 52–59
14. Chau DH, Pandit S, Wang S, Faloutsos C (2007) Parallel crawling for online social networks. In: Proceedings of the international conference on world wide web (WWW'07), Banff, Alberta. ACM, New York, pp 1283–1284
15. Cheng X, Dale C, Liu J (2008) Statistics and social network of Youtube videos. In: Proceedings of the international workshop on quality of service (IWQoS 2008), Enschede. IEEE, Los Alamitos, CA, USA, pp 229–238
16. Clauset A, Shalizi CR, Newman MEJ (2009) Power-law distributions in empirical data. SIAM Rev 51(4):661–703
17. Dai BT, Chua FCT, Lim EP (2012) Structural analysis in multi-relational social networks. In: Proceedings of the international SIAM conference on data mining (SDM 2012), Anaheim. Omnipress, Madison, pp 451–462
18. De Meo P, Ferrara E, Fiumara G, Provetti A (2011) Generalized Louvain method for community detection in large networks. In: Proceedings of the international conference on intelligent systems design and applications (ISDA 2011), Cordoba. IEEE, Los Alamitos, CA, USA, pp 88–93
19. de Sola Pool I, Kochen M (1978) Contacts and influence. Soc Netw 1:5–51
20. Freeman LC (1979) Centrality in social networks conceptual clarification. Soc Netw 1(3): 215–239
21. FriendFeed (2012). http://friendfeed.com/
22. Gathera (2012). http://www.gathera.com/
23. Ghosh R, Lerman K (2010) Predicting influential users in online social networks. In: Proceedings of the KDD international workshop on social network analysis (SNA-KDD'10), San Diego. ACM, New York
24. Google Open Social (2012). http://code.google.com/intl/it-IT/apis/opensocial/
25. Gilbert E, Karahalios K (2009) Predicting tie strength with social media. In: Proceedings of the international conference on human factors in computing systems (CHI'09), Boston. ACM, New York, pp 211–220
26. Gilbert AC, Levchenko K (2004) Compressing network graphs. In: Proceedings of the international workshop on link analysis and group detection (LinkKDD'04), Seattle. ACM, New York
27. Gjoka M, Kurant M, Butts CT, Markopoulou A (2010) Walking in Facebook: a case study of unbiased sampling of OSNs. In: Proceedings of the international conference on computer communications (INFOCOM'10), San Diego. IEEE, Los Alamitos, CA, USA, pp 1–9
28. Granovetter MS (1973) The strength of weak ties. Am J Sociol 78(6):1360–1380
29. Kahn AB (1962) Topological sorting of large networks. Commun ACM 5(11):558–562
30. Kazienko P, Musial K, Kukla E, Kajdanowicz T, Bródka P (2011) Multidimensional social network: model and analysis. In: Proceedings of the international conference on computational collective intelligence (ICCCI 2011), Gdynia. Springer, Berlin, pp 378–387
31. Kleinberg J (2008) The convergence of social and technological networks. Commun ACM 51(11):66–72
32. Korolova A, Motwani R, Nabar SU, Xu Y (2008) Link privacy in social networks. In: Proceedings of the ACM international conference on information and knowledge management (CIKM'08), Napa Valley. ACM, New York, pp 289–298

33. Krishnamurthy V, Faloutsos M, Chrobak M, Lao L, Cui JH, Percus A (2005) Reducing large internet topologies for faster simulations. In: Proceedings of the international conference on networking (Networking 2005), Waterloo, Ontario. Springer, Berlin, pp 165–172
34. Krishnamurthy B, Gill P, Arlitt M (2008) A few chirps about Twitter. In: Proceedings of the first workshop on online social networks, Seattle, pp 19–24
35. Kumar R, Novak J, Tomkins A (2010) Structure and evolution of online social networks. In: Link mining: models, algorithms, and applications, Springer, New York, pp 337–357
36. Kurant M, Markopoulou A, Thiran P (2010) On the bias of BFS (Breadth First Search). In: Proceedings of the international teletraffic congress (ITC 22), Amsterdam. IEEE, Los Alamitos, CA, USA, pp 1–8
37. Lee SH, Kim PJ, Jeong H (2006) Statistical properties of sampled networks. Phys Rev E 73(1):016102
38. Leskovec J, Faloutsos C (2006) Sampling from large graphs. In: Proceedings of the ACM SIGKDD international conference on knowledge discovery and data mining (KDD'06), Philadelphia. ACM, New York, pp 631–636
39. Li YM, Lai CY, Chen CW (2011) Discovering influencers for marketing in the blogosphere. Inf Sci 181(23):5143–5157
40. Liben-Nowell D, Novak J, Kumar R, Raghavan P, Tomkins A (2005) Geographic routing in social networks. Proc Natl Acad Sci USA 102(33):11623–11628
41. Lovász L (1993) Random walks on graphs: a survey. In: Combinatorics, Paul Erdos is eighty, vol 2, no 1, Springer, Heidelberg, Germany, pp 1–46
42. Mathioudakis M, Koudas N (2009) Efficient identification of starters and followers in social media. In: Proceedings of the international conference on extending database technology: advances in database technology (EDBT '09), Saint Petersburg. ACM, New York, pp 708–719
43. Mislove A, Marcon M, Gummadi KP, Druschel P, Bhattacharjee B (2007) Measurement and analysis of online social networks. In: Proceedings of the ACM SIGCOMM international conference on internet measurement (IMC'07), San Diego. ACM, New York, pp 29–42
44. Mislove A, Koppula HS, Gummadi KP, Druschel F, Bhattacharjee B (2008) Growth of the Flickr social network. In: Proceedings of the international workshop on online social networks (WOSN'08), Seattle. ACM, New York, pp 25–30
45. Monclar R, Tecla A, Oliveira J, de Souza JM (2009) MEK: using spatial–temporal information to improve social networks and knowledge dissemination. Inf Sci 179(15):2524–2537
46. Mucha PJ, Richardson T, Macon K, Porter MA, Onnela J (2010) Community structure in time-dependent, multiscale, and multiplex networks. Science 328(5980):876–878
47. Musiał K, Juszczyszyn K (2009) Properties of bridge nodes in social networks. In: Proceedings of the international conference on computational collective intelligence (ICCCI 2009), Wroclaw. Springer, Berlin, pp 357–364
48. Newman MEJ (2002) Assortative mixing in networks. Phys Rev Lett 89(20):208701
49. Onnela JP, Reed-Tsochas F (2010) Spontaneous emergence of social influence in online systems. Proc Natl Acad Sci 107(43):18375
50. Perer A, Shneiderman B (2006) Balancing systematic and flexible exploration of social networks. IEEE Trans Vis Comput Graph 12(5):693–700
51. Power.com (2012). http://techcrunch.com/2008/11/30/powercom-for-social-networking-power-users/
52. Rafiei D, Curial S (2005) Effectively visualizing large networks through sampling. In: Proceedings of the IEEE visualization conference 2005 (VIS'05), Minneapolis. IEEE, Los Alamitos, CA, USA, p 48
53. Rasti AH, Torkjazi M, Rejaie R, Stutzbach D (2008) Evaluating sampling techniques for large dynamic graphs. Univ. Oregon, Tech. Rep. CIS-TR-08-01
54. Romero DM, Galuba W, Asur S, Huberman BA (2011) Influence and passivity in social media. In: Proceedings of the international conference on world wide web (WWW'11), Hyderabad. ACM, New York, pp 113–114
55. Song X, Chi Y, Hino K, Tseng B (2007) Identifying opinion leaders in the blogosphere. In: Proceedings of the ACM international conference on information and knowledge management (CIKM'07), Lisbon. ACM, New York, pp 971–974

56. Stanford Network Analysis Package (2012). http://snap.stanford.edu/snap/
57. Stutzback D, Rejaie R, Duffield N, Sen S, Willinger W (2006) On unbiased sampling for unstructured peer-to-peer networks. In: Proceedings of the international conference on internet measurements, Rio De Janeiro. ACM, New York, pp 27–40
58. Travers J, Milgram S (1969) An experimental study of the small world problem. Sociometry 32(4):425–443
59. Wilson C, Boe B, Sala A, Puttaswamy KPN, Zhao BY (2009) User interactions in social networks and their implications. In: Proceedings of the ACM European conference on computer systems (EuroSys'09), Nuremberg. ACM, New York, pp 205–218
60. Wu A, DiMicco JM, Millen DR (2010) Detecting professional versus personal closeness using an enterprise social network site. In: Proceedings of the international conference on human factors in computing systems (CHI'10), Atlanta. ACM, New York, pp 1955–1964
61. XFN - XHTML Friends Network (2012). http://gmpg.org/xfn
62. Ye S, Lang J, Wu F (2010) Crawling online social graphs. In: Proceedings of the international Asia-Pacific web conference (APWeb'10), Busan. IEEE, Los Alamitos, CA, USA, pp 236–242

Privacy and Ethical Issues in Social Network Analysis

Lei Chen and Mingxuan Yuan

Abstract Nowadays, more and more people join social networks, such as Facebook, Linkedin, and Livespace, to share information and to monitor or participate in different activities. This gives people a great opportunity to obtain useful information from these social network data. Meanwhile, the information stored in these social networks are under high risk of attack by various malicious users, in other words, people's privacy could be easily breached via some domain knowledge. Thus, for a service provider, such as Facebook and Linkedin, how to publish a privacy preserving graph becomes an important problem. It is essential to protect users' privacy and at the same time provide "useful" data. Targeting the privacy preserving graph publication problem, in this chapter, we introduce graph publishing models, which cover different aspects of graph publication issues.

1 Introduction

1.1 What Is Social Network Data

With the rapid growth of Web 2.0, more and more users are joining social networks for information sharing. A social network could be modeled as a social graph where each node represents a user and the links represent the relationships between these

L. Chen (✉)
The Hong Kong University of Science and Technology, Clear Water Bay, New Territories, Hong Kong
e-mail: leichen@cse.ust.hk

M. Yuan
Hong Kong Huawei Noah's Ark Lab, Hong Kong
e-mail: yuan.mingxuan@huawei.com

Ş. Gündüz-Öğüdücü and A.Ş. Etaner-Uyar (eds.), *Social Networks: Analysis and Case Studies*, Lecture Notes in Social Networks, DOI 10.1007/978-3-7091-1797-2_9,

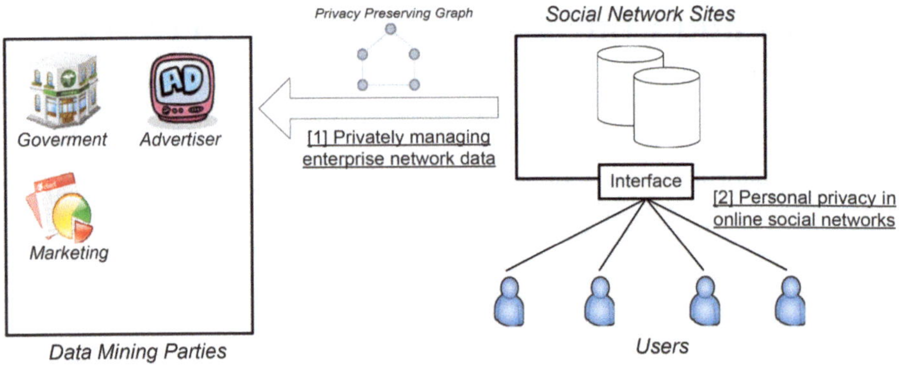

Fig. 1 Privacy-aware data management in social networks

users. Each node can have a unique identifier and a number of attributes, such as age, education background, income and so on.

It is a great opportunity to obtain useful information from these social network data since lots of information unavailable before is now searchable online [1–8]. However, social networks often contain the private information of individuals as well as the sensitive relationships between them. As a result, how to organize and manage the social networks (graphs) to make the data analysis not reveal private information has become a hot research topic.

1.2 What Is Privacy Preserving Graph Data Publishing

Figure 1 demonstrates the structure of privacy-aware data management in social networks. Users share their information through the interface provided by the social network owners. The owner holds the network data and provides the privacy preserving graph to the data mining parties to do further analysis. The privacy-aware data management in social networks involves two categories [9]:

- Privately Managing Enterprise Network Data:
  This category targets on how to publish a privacy preserving network to make sure no individual's private information is released and at the same time, the data mining parties can still get useful information from the graph.
- Personal Privacy in Online Social Networks:
  This category targets on how to guide the individual to share information or set privacy preferences when using a social network in various places. The purpose is to make the sharing of information reliable and transparent.

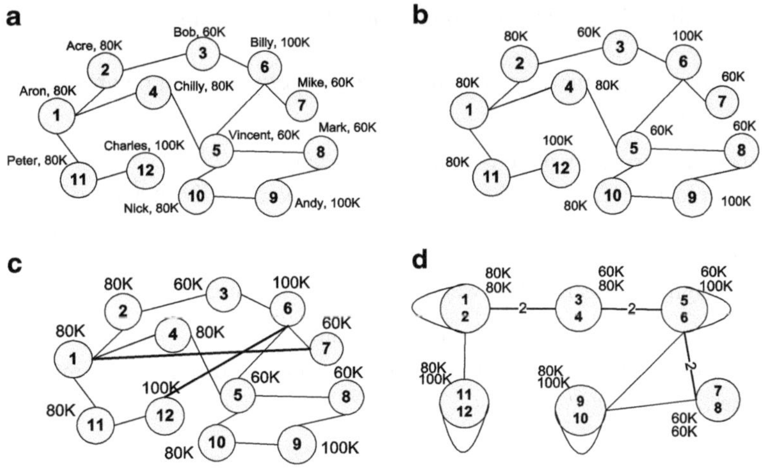

Fig. 2 Example of privacy preserving graph publication. (**a**) Original graph; (**b**) naive protection; (**c**) protection by modifying; (**d**) protection by generalizing

1.2.1 Privately Managing Enterprise Network Data

The goal is to publish a privacy preserving graph which can be used for data analysis and at the meantime, no private data can be revealed. There are four key issues in privacy preserving graph publication:

What Is the Privacy Concern

One common practice [10–12] to protect privacy is to publish a naive node-anonymized version of the network, that is, by removing the identifiers of the nodes. However, simply removing the identifiers in social networks does not guarantee privacy. The unique patterns, such as node degree or subgraph to special nodes, can be used to re-identify the nodes/links [10, 13, 14]. For example, Fig. 2a is an example of a social network where each node has a name (identifier) and income as attributes. Figure 2b is the naive node-anonymized version of Fig. 2a. If an attacker knows Aron has three friends (the node corresponding to Aron has degree 3), he can uniquely locate node 1 as Aron (we also called the process re-identification). When a node is re-identified, the information around this node in the published graph as well as the position of the corresponding node in the whole network is released. So the identities of nodes should be protected [10, 13–21]. In some applications, if the nodes have sensitive attributes, the sensitive node attributes such as income also need to be protected [14,22,23]. Some applications [14,18,21] might further require the relationship between users to be sensitive. In this case, the published graph must also guarantee no link can be re-identified. Besides the above node and link re-identification, for some applications such as the business network, a weighted graph is published [24, 25] in which only the edge weights are required to be protected.

To summarize, the privacy breaches in social networks can be grouped into four categories [11, 12]:

- Identity disclosure: A node is uniquely re-identified.
- Link disclosure: The sensitive relationship between individuals is disclosed.
- Attribute disclosure: The sensitive attribute associated with each node is compromised.
- Edge weight disclosure: The weights on edges are disclosed.

The published privacy preserving graph must avoid one or more of the above privacy breaches according to different applications and different assumptions about an attacker.

What Is the Attack Model Used to Break Users' Privacy

An attacker may obtain private information about some users using the published graph and some knowledge he already has about the users. The knowledge an attacker uses is called background knowledge. So the attack is also known as a "Background knowledge-based attack" [11]. The assumptions of the adversary's background knowledge play a critical role in modeling attacks and developing methods to protect privacy in social network data [12]. There are different types of background knowledge: non-sensitive node attributes, degrees, link relationships, neighborhoods, and embedded subgraphs, and so on. [11]. For example, knowing that Aron has three friends (the node corresponding to Aron has degree 3) to re-identify Aron is the attack using the degree background knowledge.

The assumption of background knowledge is related to the applications and graph models. Each proposed graph protection model is based on a certain assumption of background knowledge that an attacker may use. The protection techniques rely heavily on background knowledge. It is very challenging [9, 11, 12] to model all types of adversary background knowledge and quantify their impact on privacy breaches under the scenario of privacy preserving social network data publishing.

What Utilities Should Be Preserved to Guarantee Data's Usefulness

Since the main purpose of publishing a social network is to do analysis and data mining, it is important to preserve the utilities of the original graph in the published privacy preserving graph. How to measure the information loss in privacy preserving graphs is a key issue. Previous works considered three types of utilities [11, 12]:

- Graph topological properties
  Researchers used various graph topological properties as measures to indicate the structure and characteristics of social networks. These topological properties

include degree sequences, shortest connecting paths, and clustering coefficients and so on [10, 13–16].

- Aggregate queries on a network
  An aggregate network query calculates the aggregate on some paths or subgraphs satisfying some query conditions. For instance, how many users from the US are friends with users from Australia is one of the aggregate queries [21]. Some researchers used this utility to measure the labeled graphs [15, 21].
- The spectrum of a network
  Some researchers also used the spectrum of a graph to represent the structural information. The spectrum of a graph is defined as the set of eigenvalues of the graph's adjacency matrix [17, 26, 27].

Owing to the complexity of graph data, so far, how to quantify the information loss for different applications well is still an open problem [11, 12].

What Are the Fundamental Protection Techniques

To prevent the node disclosure, link disclosure and attribute disclosure, similar to the protection of tabular data, the anonymization techniques such as k-anonymity, l-diversity models are designed for graph data by considering the new attack models and utilities. There are two basic methods to publish a privacy preserving graph:

- Graph Modifying: Graph modifying is also known as the edge editing based method. The structure of the original graph is changed to satisfy a certain privacy requirement by adding/deleting edges. For instance, Fig. 2c is a privacy preserving graph of Fig. 2a by adding two edges, in which each degree appears at least twice. An attacker with the node degrees as the background knowledge, always gets two candidates for each individual. This is an implementation of 2-anonymity on node degrees by edge editing.
- Graph Generalizing: Graph generalizing is also known as the clustering based method. Clustering is to cluster "similar" nodes and publish clusters instead of the original nodes. The published graph is called a clustered graph. For instance, Fig. 2d implements 2-anonymity of nodes for Fig. 2a by clustering. Since there are many graphs which are consistent with a clustered graph, when measuring the utilities of a clustered graph, researchers sample a group of graphs which are consistent with the clustered graph and use the average utility values of these sampling graphs.

Currently, all works evaluate utilities of the published graph by empirically comparing it to the original graph. No privacy preserving graph construction algorithms generate a published graph with a bounded utility. It is an open problem about how to generate privacy preserving graphs with better utilities [12].

1.2.2 Personal Privacy in Online Social Networks

Different to the private management of enterprise network data, the works about personal privacy in online social networks focus on how to provide assistance when users share information or do the privacy preference settings through the social network interface. These works include [9]:

- Understanding your privacy risk
  The works in this category try to help users develop intuition about his information safety in online social networks. Then a user can decide whether to share the information according to the current privacy estimation.
- Managing your privacy control
  The works in this category consider how to assist a user to configure his privacy preferences more efficiently and accurately.

1.3 Content of This Chapter

In this chapter, we focus on the private management of enterprise network data. The complexity of graph data makes privacy protection more challenging than tabular data. The new challenges in this area include:

1. In terms of utility preserving
   The challenges are how to improve the utilities of the published graph, or how to generate a utility preserving graph.
2. In terms of background knowledge
   The challenges are how to model the adversary's background knowledge for different applications, how to publish a utility preserving graph against the corresponding attacks and so on.
3. In terms of attacks
   The challenges are to study whether there are additional attacks on existing privacy preserving models and so on.

After a brief introduction to related works in Sect. 2, each of the following sections tries to meet one or more challenges mentioned above.

1. Considering the utility preserving, a k-degree-l-diversity model is proposed and the corresponding utility preserving graph construction algorithm is designed in Sect. 3 [28].
2. Considering the modeling of background knowledge, a personalized protection framework for social networks is proposed in Sect. 4 [29].
3. Considering the protection against a new type of attack: analyzing attack using the label-structure relationship, a utility preserving protecting model is proposed in Sect. 5.
4. Considering the utility preserving and privacy protection, a fine-grained adjusting framework is constructed to publish the privacy protected and utility preserved graph in Sect. 6 [30].

2 Related Work

2.1 Privately Managing Enterprise Network Data

When publishing privacy preserving social networks, simply removing the identifiers does not guarantee privacy. Unique patterns, such as node degree or subgraph to special nodes, can be used to get individuals' private information [10]. An attack is implemented by using some knowledge about the victims to analyze the published graph, which is called "Background knowledge-based attacks" [11]. If the "Background knowledge" includes the structural information, it is also called "structural attacks".

If the background knowledge is the information that an attacker observed around victims, the corresponding attack is called a "passive attack". There are two models proposed to publish a privacy preserved graph against "passive attacks": the edge editing based model [13–17] and the clustering based model [10, 18–21]. The edge editing based model is to add or delete edges to make the graph satisfy certain properties according to the privacy requirements. The clustering based model is to cluster "similar" nodes together to form super nodes. Each super node represents several nodes which are also called a "cluster". Then the links between nodes are represented as the edges between super nodes which are called "super edges".

Most edge editing based graph protection models implement k-anonymity [31] of nodes on different assumptions of the background knowledge of the attacker. That is, when an attacker uses certain background knowledge to query the published graph, he should always get at least k candidates. By assuming that an attacker knows the degree of each node, Liu [13] did a pioneer work in this area. Liu [13] defined and implemented the k-degree-anonymous model on network structure, that is for a published network, for any node, there exists at least other $k - 1$ nodes to have the same degree as this node. The published graph is generated by changing the original graph with the minimum number of edges. Similarly, several other works propose different k-anonymous graph protection models with stronger assumption of the attacker's background knowledge. Zhou [15] considered the k-neighborhood anonymous model: for every node, there exists at least other $k - 1$ nodes sharing isomorphic neighborhoods. They assume that an attacker knows the one hop neighborhood graph of each node. In paper [22], the k-neighborhood anonymity model is extended to k-neighborhood-l-diversity model to protect the sensitive node label. That is, the sensitive labels on the nodes who share the isomorphic neighborhoods also satisfies l-diversity [32]. Zou [16] made the strongest assumption where an attacker can know arbitrary subgraph around a node. Then they proposed a k-Automorphism protection model: A graph is k-Automorphism if and only if for every node there exist at least $k - 1$ other nodes which do not have any structural difference with it. Cheng [14] also assumed that an attacker can know arbitrary subgraph around a node. They implemented the k-anonymity on both nodes and links. The k-isomorphism model proposed by them to protect both nodes and links is: a graph is k-isomorphism if it consists of k disjoint isomorphic subgraphs.

Besides k-anonymity, several other works published a randomly changed graph with edge editing. Since the published graph has been randomly changed, the connection information around users is protected through the random change. Ying [17] proposed a protection model which randomly changes the edges in the graph. They studied how randomly deleting and swapping edges changes graph properties and proposed an eigenvalues oriented random graph change algorithm. To show the privacy preserving effect of random edge change, Bonchi [33] proposed an entropy-based measure which can quantify random changed graph's anonymity level. Xiao [34] proposed a randomization scheme, which obfuscates edge existence of graphs to protect both nodes and edges.

Besides edge editing, the other publication method is clustering. Since a clustered graph only contains super nodes and super edges, by making each cluster's size at least k, the probability of re-identifying a user can be bounded to $\frac{1}{k}$. Hay [10] proposed a heuristic clustering algorithm to prevent privacy leakage using vertex refinement, subgraph, and hub-print attacks. Campan [19] discussed how to implement clustering when considering the loss of both node labels and structural information. Zheleva [18] developed a clustering method to prevent the sensitive link leakage. Cormode [20, 21] introduced (k,l)-clusterings for bipartite graphs and interaction graphs respectively. Compan [23] implemented a p-sensitive-k-anonymity clustering model which requires each cluster to satisfy distinct l-diversity [32].

There are also two other works which focus on the edge weight protection in weighted graphs. They only protect the weights on edges instead of nodes, links and node attributes. The attacking scenario is that an attacker wants to learn the true values of edge weights or the relative order between two edge weights by observing the published weighted graph. The protection method is to replace all edge weights with new values. Then the true values of edge weights or the relative order between two edge weights are hidden. Liu [24] proposed a random edge weight assignment method to preserve the shortest paths between most pairs of nodes in the graph. Das [25] proposed a Linear Programming based edge weight assignment method to protect the edge weights while preserving certain linear properties of the graph.

Besides the "passive attack", there's another type of attack on social networks, which is called an "active attack". An "active attack" is to actively embed special subgraphs into a social network when it is collecting data. An attacker can attack the users who are connected with the embedded subgraphs by re-identifying these special subgraphs in the published graph. The background knowledge an attacker uses is the subgraph he intentionally created instead of the information he observed. Backstrom [35] described active attacks based on randomness analysis and demonstrated that an attacker may plant some constructed substructures associated with the target entities. A "passive attack" only works on social networks which do not need double authentification to build a link (a link can only be built when both users agreed to build it), such as an email network. For a social network where links can be freely added by an attacker, the method used to prevent an active attack is to recognize the fake nodes added by attackers and remove them before publishing the data. Shrivastava [36] and Ying [27] focused on a special active attack named

Random Link Attack. Shrivastava [36] proposed an algorithm that can identify fake nodes based on the triangle probability difference between normal nodes and fake nodes. Ying [27] proposed another method, which uses spectrum analysis to find the fake nodes. To publish a graph that is potentially changed by an active attack, the publisher can use a two-step mechanism. Firstly, the graph is filtered by the methods introduced by Backstrom [35] or Shrivastava [36]. Then the published graph can be generated using protection models for the passive attack from the filtered graph.

Besides the "passive attack" and "active attack", Arvind [37] studied the ability of an attacker to re-identify nodes in an anonymized graph G_a when he knows a different graph G_{aux} whose membership partially overlaps with G_a. The attacker knows the mapping of some "seed nodes" between G_a and G_{aux}. Arvind [37] showed that from these "seed nodes", a large portion of other nodes can be re-identified.

2.2 Personal Privacy in Online Social Networks

Supporting tools should be developed to guide the operations of users to avoid the leakage of private information when people share information or do the privacy preference settings in an online social network.

Liu [38] proposed a framework to compute a privacy score of a user, which indicates the potential privacy risk caused by his participation in the network. Some works analyzed the ability of an attacker to learn the unpublished attributes of users in an online social network when he uses the published attributes and relationships between users to do the data mining. They showed that individual's private information can be inferred using majority voting [39, 40], community detection [41] and classification techniques [40, 42].

Fang [43] designed a wizard which helps a user to do the privacy preference setting using a limited amount of user input. A machine learning model, which uses the privacy settings on a small number of friends as a training set to automatically assign privileges to other users, is designed. Shehab [44] also proposed an approach that assists users in composing and managing their access control policies. They designed a supervised learning mechanism that leverages user provided example policy settings as training sets to build classifiers.

3 Protecting Sensitive Labels in Social Network Data Anonymization

3.1 Motivation

Liu et al. [13] did pioneer work in privacy preserving graph publication. He defined a k-degree anonymity model to prevent degree attacks (Attacks use the degree knowledge of a node). A graph is k-degree anonymous if and only if for any node in this graph, there exist at least $k - 1$ other nodes with the same degree. However, there are two shortcomings with this model. The first one is that if the nodes have sensitive labels, k-anonymity cannot protect these sensitive information well. In many applications, a social network where each node has sensitive attributes should be published [22, 23]. The advanced models such as l-diversity must be considered. Again, the l-diversity concept here has the same meaning as that defined over tabular data [32]. For example, if we choose the distinct l-diversity, for nodes with same degree, their associated sensitive labels must have l distinct values. Another shortcoming is that generating the k-degree anonymous graph by edge editing may largely destroy the properties of a graph. The edge editing method sometimes may change the distance properties substantially by connecting two faraway nodes together or deleting the bridge link between two communities. Mining over such data might get the wrong conclusion about, for example, how the salaries are distributed in the society.

To address these issues, we firstly extend the k-degree anonymous model to k-degree-l-diversity model to protect sensitive labels. Then, we propose a novel idea to preserve important graph properties, such as distances between nodes by adding certain "noise" nodes into a graph. This idea is based on the following key observation. Most social networks satisfy the Power Law distribution [45], there exist a large number of low degree vertices in the graph which could be used to hide added noise nodes from being re-identified. By carefully inserting noise nodes, some graph properties could be better preserved than a pure edge editing method.

In this section, we consider a graph model where each vertex in the graph is associated with a sensitive label. The privacy preserving goal is to prevent an attacker from finding the fact that a certain user has a specific sensitive value. To achieve this goal, we define a k-degree-l-diversity model for safely publishing a labeled graph, and then develop the corresponding graph anonymization algorithms with the least distortion to the properties of the original graph, such as degrees and distances between nodes. The discussion about how to adapt this technique into more complicated cases can be found in [28]

3.2 Problem Description

In this section, a social network graph is defined as:

Definition 1. Social Network Graph: a social network graph is a four tuple $G(V, E, \sigma, \lambda)$, where V is a set of vertices, and each vertex represents a node in the social network. $E \subseteq V \times V$ is the set of edges between vertices, σ is a set of labels that vertices have. $\lambda : V \longrightarrow \sigma$ maps vertices to their labels.

In the rest part, we use the words "node" and "vertex" interchangeably. In a published (privacy preserving) social network graph, an attacker could re-identify a node by degree information [13] and further infer sensitive labels. To prevent this possible leakage, we define the "k-degree-l-diversity" principle for published graphs, which have the same spirit of $k - l$ diversity in relational data [31, 32].

Definition 2. k-degree-l-diversity (*KDLD*): For each vertex in a graph, there exists at least $k - 1$ other vertices having the same degree in the graph. Moreover, the vertices with the same degree contain at least l distinct sensitive labels.

Here, we use distinct l-diversity [32] to ensure that there exists at least l distinct labels in each equivalent class (group), i.e., a group of nodes having the same degree.[1] We use distinct l-diversity to demonstrate the basic working procedure of our method. The detailed discussion about how to handle more complex l-diversity such as recursive (c, l)-diversity can be found in [28]. We call a graph a *KDLD graph* if it satisfies the k-degree-l-diversity constraint. A *KDLD graph* protects two aspects of each user when an attacker uses degree information to attack: (1) The probability that an attacker can correctly re-identify this user is at most $\frac{1}{k}$; (2) The sensitive label of this user can at least be related with l different values. Since each equivalent class contains at least k nodes, when an attacker uses the degree to re-identify a node, the probability he correctly re-identifies this user is at most $\frac{1}{k}$. Furthermore, since there are at least l distinct labels in each equivalent class, a user's sensitive label is related with at least l values. Our goal is to protect each user's privacy by adding certain edges/nodes to transfer a social network graph to a KDLD graph. We call the added edges/nodes noise edges/nodes. To guarantee the correctness of information, for any user who appears in the published graph, we also preserve the sensitive label associated with this user's node. In order to keep the utility of the published graph, when generating the *KDLD* graph, two key properties should be preserved [46]:

- Low Overhead: It is necessary to add as few noise edges as possible to reduce the additional overhead on the social infrastructure;
- Social Distance: The noise edges/nodes added should connect nodes that are close with respect to the *social distance.*

Since each noise node connects with at least one noise edge, "Low Overhead" also limits the number of noise nodes that can be added. The social distance between two nodes u, v is the shortest path length between u and v in the original graph. The social distance between all node pairs of a graph is measured by **Average Shortest Path Length** (*APL*). *APL* is a concept in network topology that is defined as the average of the distances between all pairs of nodes. It is a measure of the efficiency of information or mass transport on a network. Some queries like "the nearest node for a group of nodes" are related with *APL*. The *APL* of a graph G is $APL_G = \frac{2}{N(N-1)} \sum_{\forall n_i, n_j \in G} d(n_i, n_j)$. In it, $d(n_i, n_j)$ is the length of the shortest path between nodes n_i and n_j, N is the number of nodes in the graph.

[1] If nodes have other labels (quasi-identifiers) than the sensitive label, the quasi-identifiers of the nodes in each equivalent class will be adjusted to be the same.

We design a two step algorithm to generate the *KDLD* graph which tries to preserve the above two key properties. In the first step, we compute a target degree for each node so that it makes the original graph satisfy *KDLD* constraint with the minimum sum of degree change. Clearly smaller degree change needs fewer noise edges to implement the change. In the second step, we change each node's degree to its target degree by adding noise edges/nodes. We utilize the noise nodes to make the change of *APL* as small as possible. Our proposed two step algorithm considers both the "Low Overhead" and the "Preserve Social Distance" requirements.

Next, we first introduce two data structures we use in the rest of this section, and then give the formal description the two steps in our algorithm. We borrow the concept of "degree sequence" used in [13] and define a *sensitive degree sequence P* as follows:

Definition 3. Given a graph G, its sensitive degree sequence is a sequence of n triples: $[P[1], \ldots, P[n]]$ where $P[1].d \geq P[2].d \geq \ldots \geq P[n].d$, $P[i]$ is a triple (id, d, s) at position i in P, d is the degree, and s is the sensitive label associated with id.

We use lowercase p_u to represent node u's corresponding triple in P. For a node u, we use $u.d$ to denote its degree and $u.s$ for its label. Now, we can define *KDLD* model of a graph based on its sensitive degree sequence:

Definition 4. *KDLD* sequence: A sensitive degree sequence P is a *KDLD* sequence if P satisfies the following constraint: P can be divided into a group of subsequences $[[P[1], \ldots, P[i_1]], [P[i_1 + 1], \ldots, P[i_2]], [P[i_2 + 1], \ldots, P[i_3]], \ldots, [[P[i_m + 1], \ldots, P[j]]$ such that for any subsequence $P_x = [P[i_x], \ldots, P[i_{x+1}]]$, P_x satisfies three constraints: (1) All the elements in P_x share the same degree ($P[i_x].d = P[i_x + 1].d = \ldots = P[i_{x+1}].d$); (2) P_x has size at least k ($i_{x+1} - i_x + 1 \geq k$); (3) P_x's label set $\{P[t].s | i_x \leq t \leq i_{x+1}\}$ have at least l distinct values.

For a *KDLD* sequence P, each its subsequence $[P[i_x], \ldots, P[i_{x+1}]]$ forms a same degree group c_x. Then from P, we can get same degree groups: $C = \{c_1, c_2, \ldots, c_m\}$. For each $c_i \in C$, we use $c_i.d$ to denote the degree of the corresponding group. It is obvious that a graph $G(V, E)$ is a *KDLD* graph if and only if the sensitive degree sequence of G is a *KDLD* sequence.

- *KDLD* sequence generation: Given the sensitive degree sequence P of graph G and two integers k and l, compute a *KDLD* sequence P^{new} which contains the same set of nodes as P with minimum $L(P, P^{new}) = \sum_{\forall u} |p_u.d - p_u^{new}.d|$. This equation computes the degree change by comparing the same node u's degree in these two sequences. The purpose is to obtain a new *KDLD* sequence from G so that the degree change of all the nodes in G is as small as possible. Clearly smaller degree change needs fewer noise edges to implement the change.
- Graph Construction: Given a graph $G(V, E, \sigma, \lambda)$ and a sensitive degree sequence P^{new}, construct a new graph $G'(V', E', \sigma, \lambda')$ with $V \subseteq V'$. The sensitive degree sequence P' of G' is a *KDLD* sequence and P' has all the elements in P^{new}

since G' is constructed from G by adding some noise nodes. Meanwhile $|APL_G - APL_{G'}|$ is minimized.

Next, we introduce how to implement these two steps.

3.3 KDLD Sequence Generation

To generate a *KDLD* sequence, the triples in P should be divided into groups. All the corresponding nodes in the same group shall be adjusted to have the same degree. We employ two algorithms for this problem: Algorithm K-L-BASED and Algorithm L-K-BASED.[2] The algorithms tend to put the nodes with similar degrees into the same group to reduce the degree changes. Given the degree sequence P of the original graph, Algorithm K-L-BASED selects the first k elements in P as a group. We keep on merging the next element into the current group until the l-diversity constraint is satisfied. After a group satisfies k-degree and l-diversity constraints, we calculate two costs:

- C_{new}: the cost of creating a new group for the next k elements (the total degree changes that make all the nodes in this group have the same degree).
- C_{merge}: the cost of merging the next element into the current group and creating a new group for the next k elements by skipping the next element.

If C_{merge} is smaller, we merge the next element into the current group and continue this comparison process. If C_{new} is smaller, we create a new group with the next k elements and continue checking the l-diversity constraints described above. When the remaining elements are less than k or contain less than l distinct sensitive label, they are merged into the last group. Since nodes are sorted by their degrees in P, building groups using the above method helps to group the nodes with similar degrees together.

Different from Algorithm K-L-BASED which checks the k-degree first, Algorithm L-K-BASED tries to satisfy the *l*-diversity first. If we use t to represent the position number of each element in P. Each time Algorithm L-K-BASED picks l non-grouped *P[t]*s in P that do not share any common labels with the minimum $\sum t$. Minimum $\sum t$ guarantees the algorithm select l nodes with most similar degrees in the non-grouped nodes. The algorithm then continues merging *P[t]* with minimum t in the remaining elements until we get k items. Then similar to Algorithm K-L-BASED, we use the same cost function to determine if the next element should be merged or not.

All the nodes in the same group will be adjusted to have the same degree in the next graph construction step. We use the mean degree of a group as its target

---

[2]For the case that each node has some quasi-identifier labels, the generalization cost on quasi-identifier labels in each group can be combined into the cost function in the *KDLD* sequence generation algorithms.

degree. This target degree is also used to estimate the cost in Algorithm K-L-BASED and Algorithm L-K-BASED.

3.4 Graph Construction

3.4.1 Algorithm Skeleton

After getting the new sensitive degree sequence P^{new}, a new graph G' should be constructed. Suppose P' is the sensitive degree sequence of G', P'_o is the sensitive degree sequence of G' that only contains the nodes in G. Our graph construction algorithm makes $P'_o == P^{new}$ and P' as a *KDLD* sequence.

We use the following algorithm to construct the published graph which preserves the *APL*. The algorithm contains five steps:

- Step 1: Neighborhood_Edge_Editing()
  We add or delete some edges if the corresponding edge editing operation follows the ***neighborhood rule*** (The details can be found in Step 1). By doing this, the sensitive degree sequence P of original graph G is closer to P^{new} in case *APL* is preserved;
- Step 2: Adding_Node_Decrease_Degree()
  For any node whose degree is larger than its target degree in P^{new}, we decrease its degree to the target degree by making use of noise nodes;
- Step 3: Adding_Node_Increase_Degree()
  For any node whose degree is smaller than its target degree in P^{new}, we increase its degree to the target degree by making use of noise nodes;
- Step 4: New_Node_Degree_Setting()
  For any noise node, if its degree does not appear in P^{new}, we do some adjustment until it has a degree in P^{new}. Then the noise nodes are added into the same degree groups in P^{new};
- Step 5: New_Node_Label_Setting()
  We assign sensitive labels to noise nodes to make sure all the same degree groups still satisfy the requirement of the distinct l-diversity. It is obvious that after Step 4 and Step 5, the sensitive degree sequence P' of the published graph G' is a KDLD sequence.

In Step 2, Step 3 and Step 4, we carefully connect the noise nodes into the graph to make as little change to *APL* as possible. Next, we introduce these five steps in detail.

Step 1

We use Algorithm 1 to change some nodes' degrees by adding/deleting edges first. In order to preserve *APL*, we force each change to follow the ***neighborhood rule*** described as follows:

Algorithm 1: Neighborhood_Edge_Editing()

```
for each node u to increase degree do                              /* Case 1 */
    d = u's degree;
    d' = u's target degree;
    for i = 0; i < d' − d; i++ do
        Find v,w has link (u, v), (v, w) and v needs to decrease degree;
        if Such v,w exist then
            Remove link (v, w);
            Add link (u, w);
        else
            break;
for each node u to increase degree do                              /* Case 2 */
    for each node v need to increase degree do
        if u, v are 2 hop neighbor then
            if u, v do not have link then
                Add link (u, v);
for each node u to decrease degree do                              /* Case 3 */
    for each node v to decrease degree do
        if u, v have link then
            Remove link (u, v);
            if ¬(u, v are 2 hop neighbor) then
                Add link (u, v);
```

- Case 1: Node u needs to increase its degree, v needs to decrease its degree and u and v are direct neighbors. In this case, we randomly select one direct neighbor w of v that does not connect to u. Remove the edge (v, w) and add a new edge (u, w). By doing this, u's degree is increased by 1 and v's degree is decreased by 1 while all the other nodes' degrees are unchanged. This operation changes the distance between u, w from 2 to 1 and the distance between v, w from 1 to 2. Therefore, only the paths passing through u, w or v, w change their lengths at most by 1;
- Case 2: Two nodes u, v are 2 hop neighbors and they both need to increase their degree. If no link (u, v) exists, we add link (u, v). The distance between u, v is changed from 2 to 1. Only the lengths of the shortest paths passing through u, and v change by 1;
- Case 3: Two nodes u and v that both need to decrease their degrees are direct neighbors. If after removing link (u, v), u and v are still 2 hop neighbors, remove the link (u, v). The distance between u, v changes from 1 to 2. So it only changes the lengths of shortest paths passing through u, v at most by 1.

A node's degree could be either increased or decreased. So the above policies consider all the cases for two nodes u and v whose degrees need to be changed. It is easy to see that any edge editing operation between these two nodes may cause the lengths of shortest paths passing through u and v change at least by 2 except in the above three situations.

Step 2

For any node u whose degree is still larger than its target degree after Step 1, we decrease its degree following the steps below (Algorithm 2):

Algorithm 2: Adding_Node_Decrease_Degree()

```
for every node u that need to decrease the degree do
    d = u's degree;
    target = the target degree of u;
    while true do
        Select a sensitive value s from u's original one hop neighbor;
        Create a new node n with sensitive value s;
        d' = 1;
        target_new = Select_Closest_Degree_In_Group(d + 2 - (the target degree));
        Connect node u with n;
        d = d + 1;
        while true do
            Random select a link (u, v) which is in G;
            Delete link (u, v), Create link(n, v);
            d' = d' + 1;
            d = d - 1;
            if d' == target_new ∨ d == target then
                break;
        if d == target then
            break;
```

- Create a new node n and connect u with n, now $u.d = p_u.d + 1$ and $n.d = 1$.
- Set n's target degree $target_{new}$ to a degree in P'. If $u.d + 2 - p'_u.d$ is less than the minimum degree in P', set $target_{new}$ as this minimum degree. Otherwise, set $target_{new}$ as a degree in P' which is less than and closest to $u.d + 2 - p'_u.d$. We select $u.d + 2 - p'_u.d$ because we change u's degree through n. As a consequence, n's degree is changed to $u.d + 2 - p'_u.d$.
- Randomly select an edge (u, v) from the original graph, delete this edge, then add a new edge (n, v). By doing this, v only changes from u's 1 hop neighbor to its 2 hop neighbor. We repeat this random edge modification process until $n.d = target_{new}$ or $u.d = p'_u.d$. If $n.d = target_{new}$ and $u.d > p'_u.d$, go to the first step to create a new noise node. If $u.d = p'_u.d$ but $n.d < target_{new}$, node u's degree has been successfully adjusted to $p'_u.d$. We finish the adjustment of node u, and leave node n to noise hiding step. Since each time, we reduce u's degree

by 1 to increase n's degree by 1, and the creation of n (described in the previous step) adds degree 1 to both u and n, $u.d + 2 - p'_u.d$ is n's degree when u reaches its target degree.

Step 3

Algorithm 3: Adding_Node_Increase_Degree()

```
for every node u in G which needs to increase its degree do
    for i=0; i < increase_num; i++ do
        Create a new node n;
        Connect node u with n;
        for every node v that is one or two hop neighbor of node u do
            if v needs to increase its degree then
                Connect node v with n;
        while n's degree is not in P' ∧n's degree > min group degree do
            Remove the last connection created to n;
            i=i-1;
```

For each vertex u in G which needs to increase its degree, we use Algorithm 3 to make its degree reach the target degree. We first check whether there exists a node v within 2 hops of u, and v also needs to increase its degree (In Step 2, some noise nodes are added, such v may exist.), we connect n with v. Since v is within 2 hops of u, connecting v with n will not change the distance between u and v. After this step, if n's degree is bigger than the minimum degree in P^{new} but does not appear in P^{new}, we recursively delete the last created link until the degree of n equals to a degree in P^{new}. Otherwise, we leave n to Step 4 for processing and continue adding noise to u if $u.d < p'_u.d$. By doing this, u's degree is increased to its target degree.

Step 4

In this step, we make sure each noise node has a degree in P^{new}. By doing this, all the noise nodes are added into the existing same degree groups in P^{new}. Furthermore, since we assume an attacker uses the degree information of some nodes to do the attack, if the noise nodes have degrees as the same as some original nodes, an attacker cannot distinguish them from the original nodes with the degree information. So, we need to make all the noise nodes' degrees to be some degrees that already exist in P^{new}. To keep the *APL*, during the process, we only change connections in the noise nodes' neighborhoods to minimize the changes in distance.

The details of Step 4 are shown in Algorithm 4. We first select pairs of noise nodes (which need to increase their degrees) that each pair of nodes are within 3 hops to each other and build a link for each pair. If two noise nodes are within 3 hops, they are either connected to the same node in the original graph or they are respectively connected to two directly connected nodes in the original graph. If the two noise nodes are connected to the same original node, connecting them does not change any shortest path between the original nodes. If the two noise nodes are respectively connected to two original nodes which are direct neighbors, since connecting these two noise nodes does not change the distance between the two original nodes, the shortest path lengths between the nodes in the original graph do not change, either.

Algorithm 4: New_Node_Degree_Setting()

Select pairs of noise nodes that each pair of nodes are within 3 hops to each other;
Build a link for each pair;
for *every node n has even degree* **do**
  Select an even degree $target_{new}$ $(n.d < target_{new})$ in P^{new};
for *every node n has odd degree* **do**
  Select an odd degree $target_{new}$ $(n.d < target_{new})$ in P^{new};
for *each noise node n* **do**
  while $n.d \neq target_{new}$ **do**
    Find a link (u, v) with the minimum $\frac{dis(u,n)+dis(v,n)}{2}$ in current graph where u and v are not connected to n;
    Remove link (u, v);
    Add link (n, u);
    Add link (n, v);

Then, for each noise node whose degree is an even number, we select an even degree in P^{new} (a degree in P^{new} which is an even number) that is close to and bigger than this noise node's current degree as its target degree. For example, if a noise node's degree is 8 and the degrees appear in P^{new} are $\{3, 4, 5, 6, 7, 9, 10, 12, 15, 16\}$, we set its target degree as 10. We find all the nearest edges which do not directly connect with the current processing noise node. The "nearest" edge to a noise node means the average distance of the two points (u, v) of an edge to this noise node $\frac{dis(u,noise)+dis(v,noise)}{2}$ is minimum among all edges. We select one edge within the nearest edge set randomly, remove the edge from the graph and connect the endpoints of this edge to the current processing noise node. After that, the degree of the noise node increases by 2, meanwhile all the other vertices' degrees remain unchanged. We repeat this procedure until the noise node's degree reaches the target degree. For each noise node of an odd degree, we select an odd target degree and conduct the same adjustment.

Since both endpoints of the removed edge are near to the noise node, directly connecting them to the noise node makes them stay in the neighborhood of the noise node. Moreover, since the distance between the two endpoints increases by 1 only, the lengths of the paths through the two endpoints are increased by 1 only as well. When only even or odd degree groups exist, it is easy to change the KDLD sequence generation algorithm to get a new sensitive degree sequence that contains both even and odd degrees by increasing some groups' degrees by 1.

Step 5

The last step is to assign sensitive labels to the noise nodes to ensure the same degree groups still satisfy the requirements of distinct l-diversity. Since in each same degree group, there are already l distinct sensitive labels, it is obvious the new added noise nodes can have any sensitive label. We use the following way to assign a sensitive label to a noise node n: suppose u is the original node in G for which n is created. We randomly find a label from the direct neighbors of u in the original graph G. Since there is a node with this label directly connected to u in the original graph, the noise node n with the same label connected to u can help preserve the distances between this label and other labels in the graph.

3.5 Anonymization Utility Analysis

In Step 1, we add or delete some edges following the "neighborhood rule". To show the effectiveness of the noise nodes adding strategies in Step 2, Step 3 and Step 4, we analyze the bound of the *APL* change when given the number of added noise nodes in the average case.

Theorem 1. *Suppose G_0 is the graph generated after Step 1 with N nodes, whose APL is APL_{G_0}. If G_M is the KDLD graph generated by adding M noise nodes. Our algorithm guarantees that the APL of G_M, APL_{G_M} is bounded on average case:*

$$(1 - \frac{4}{N+1})^M APL_{G_0} \leq APL_{G_M} \leq (1 + \frac{4}{N})^M APL_{G_0} + \frac{2}{N-4}$$

The detailed proof can be found in [28].

3.6 Summary

In this section, we propose a k-degree-l-diversity model for privacy preserving graph data publishing. We design a noise node adding algorithm to construct a new graph from the original graph with the constraint of introducing fewer distortions to the original graph. We give a rigorous analysis of the theoretical bounds on the number of noise nodes added and their impacts on an important graph property.

4 Personalized Privacy Protection in Social Networks

4.1 Motivation

The previous works have overlooked a very important fact, that is, different users may have different privacy preferences. Therefore, providing the same level of privacy protection to all users may not be fair and may also render the published social network data useless.

To address the shortcomings of single level privacy protection, in this work, we define different levels of protection for users and incorporate them into the same published social network. In fact, current social network websites allow each user to decide how much personal information can be observed by others. For example, in Facebook, a user can set what part of his profile or his connection information can be viewed by others. A user can have a clear estimation of the knowledge that an attacker can know about him. The knowledge an attacker uses to find the privacy information of a user is called the background knowledge. To provide different levels of privacy protection, we allow users to set personalized privacy requirements based on their own assumptions about the attacker's background knowledge. Specifically, for a node u in a published labeled graph, starting from the weakest background knowledge that an attacker only knows u's label information without any structural information, we define three levels of attacks to u by gradually increasing the strength of the attacker's background knowledge:

1. An attacker only knows u's labels. For example, an attacker knows Bob is a 26-year old guy;
2. An attacker knows u's labels and degree. For example, an attacker knows Bob is a 26-year old guy with degree 3;
3. An attacker knows u's labels, degree and the labels on the edges adjacent to u. For example, an attacker knows Bob is a 26-year old guy with degree 3 and Bob's three connections' types are classmate, roommate, roommate;

We extract these three levels of background knowledge due to the fact that the three kinds of settings are also supported by Facebook. We noticed that there exist much stronger attacks such as knowing the neighborhood of u [11] with label information. However, in this work, our focus is to demonstrate a framework which can publish a social network to satisfy different levels of privacy protection requirements, thus, we will not enumerate all the possible attacks. In the rest part of this section, we use "Level x's background knowledge" to represent the corresponding background knowledge used in Level x's attack.

To achieve personalized privacy protection in social networks, in this work, we design different methods for different privacy requirements. Specifically, For Level 1 protection, we use node label generalization. For Level 2 protection, we combine the noise node/edge adding methods based on the protection at Level 1. For Level 3 protection, we further use the edge label generalization to achieve the protection objectives.

4.2 Problem Definition

In this section, we focus on the privacy preserving problem for an un-weighted graph with labels on both nodes and edges. The graph model used in this section is defined as follows:

Definition 5. Published Graph: a published graph for a social network is defined as a six-tuple $G(V, E, \sigma, \lambda, \beta, \lambda')$, where V is a set of nodes and each node represents a node in a social network. $E \subseteq V \times V$ is a set of edges between nodes. σ is an array, each of its dimensions represents a set of labels that nodes have. λ is an array that $\forall i, \lambda[i] : V \longrightarrow \sigma[i]$ maps nodes to their labels. β is a set of labels that edges have. $\lambda' : E \longrightarrow \beta$ maps edges to their labels.

Each node in the graph has several labels, which represent the attributes of the node. Each edge in the graph has one label, which represents the type of the edge. Since one node may have several labels, we call the labels on node u u's *label list*. We use $G(V,E)$ to simply represent the original graph where V stands for the node set and E stands for the edge set. We design a framework that allows users to set three different privacy protection requirements introduced in Sect. 4.1. The protection objectives guaranteed by this framework are: Given a constant k,

1. For each node u, the probability that an attacker re-identifies u is at most $\frac{1}{k}$.
2. For any edge e in the published graph, the probability that an attacker identifies a node u_x involved in e is at most $\frac{1}{k}$;
3. For any two nodes u_x and u_y, the probability that an attacker identifies these two nodes having a connection is at most $\frac{1}{k}$.

Theorem 2. *Suppose* cost $= \sum$ node label generalization cost $+ \sum$*edge label generalization cost. For a graph G, the problem to create a graph G_A that satisfies each user's personal privacy requirement with minimum cost is a NP-hard problem.*[3]

[3]Here the cost stands for the sum of label generalization levels. For example, in Fig. 3d, if let the labels in the group {a, d} to be the same, they become [(Africa, American), 2*], thus the node label generalization cost of this group is 4.

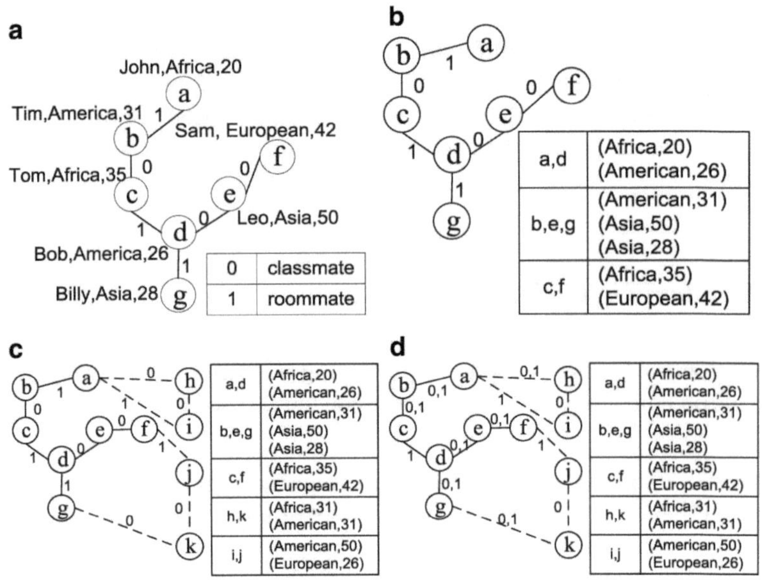

Fig. 3 Naive protection. (**a**) Original graph; (**b**) Level 1 protection; (**c**) Level 2 protection; (**d**) Level 3 protection

The proof of can be found in [29]. The three levels of background knowledge are defined from the weakest to the strongest, a user who needs a higher level protection by default needs the lower level protections. So, in our framework, we implement the personalized protections from the lowest level to the highest level. In the rest part, we use set_{l_1} to represent the set of nodes that needs Level 1 or stronger protection, set_{l_2} to represent the set of nodes that needs Level 2 or stronger protection, and set_{l_3} to denote the set of nodes that needs Level 3 protection.

4.3 Level 1 Protection

The Level 1's background knowledge is about node label list, thus, we can use a generalization method to provide Level 1 protection. Simply speaking, generalization is to group nodes and make the nodes within each group have one generalized node label list. In order to keep the node labels distribution of each group in the published graph unchanged, we do not generalize all the label lists to be the same for a group of nodes. Instead, for each group, we publish all the label lists of the nodes in the group without the mapping.[4] Figure 3b gives an example of the published graph

[4]The generalization cost can still be computed as the sum of label generalization levels when making all the labels of the nodes in each group to be the same.

after generalization the graph in Fig. 3a. This method is the same as the Anatomy for tabular data proposed by Xiao et al. [47] and has the same effect as the full list pattern permutation proposed by Cormode et al. [21].

In order to achieve Level 1 protection, we need to divide all nodes into groups and these groups must guarantee the three privacy objectives are satisfied. As shown in [21], if each group's size is at least k, the first objective is achieved. To satisfy objectives (2) and (3), the following condition [21] must be satisfied:

Theorem 3. *A division of nodes V into groups satisfies objectives (2) and (3) if for any node $u \in V$ and any group $g \subset V$:*

- $\forall (u, w), (u, z) \in E : w \in g \wedge z \in g \Rightarrow z = w.$
- $\forall (u, w) \in E : u \in g \wedge w \in g \Rightarrow u = w.$

The first condition constrains that any two nodes in the same group do not connect to a same node. The second condition constrains that no edges within a group. Theorem 3 ensures that the two objectives (2) and (3) are satisfied, but it is too restrictive. The following Theorem states that we can obtain the groups that satisfy objectives (2) and (3) easily by relaxing the constraints.

Theorem 4. *Objectives (2) and (3) in Level 1 can be guaranteed, if*

- $\forall (v, w) \in E : v \in g \wedge w \in g \Rightarrow v = w.$
- $\forall$ *group g_x, g_y, x is the number of edges between g_x and g_y, $x \leq \frac{|g_x||g_y|}{k}$.*

The first condition constrains no edge within a group and the second condition constrains the number of edges between any two groups. The proof can be found in [29]. The conditions in Theorem 4 are less restrictive than those in Theorem 3, which depend on the sparse property of social networks. As a consequence, groups are easier to be found based on the conditions in Theorem 4. For example, for Fig. 3a, if using the conditions in Theorem 3, no solution can be found. If using the conditions in Theorem 4, Fig. 3b can be constructed. We call the two conditions in Theorem 4 "***the Safety Grouping Condition (SGC)***".

Algorithm 5 computes the groups with size at least k. The algorithm targets on generating the groups under *SGC* with the minimum cost defined in Theorem 2. This algorithm has a similar structure as the algorithms in [19, 21] which also group a graph's nodes with certain cost function. Each time, we select a node with the maximum degree that has not been grouped and create a new group for this node (line 2–5). Then in line 6–18, we repeatedly add nodes into the group with the minimum cost (estimated) under *SGC* until the group reaches size k. For any group which cannot reach size k, in line 19–23, we first delete this group. Then for each node in the group, we add it into a suitable group which has the minimum cost for the insertion operation.

Finally, Algorithm 5 gets the node group set C. Then, based on C, a graph like Fig. 3b can be generated, which offers Level 1 protection for all the nodes. We use G_{L_1} to represent the generated graph at Level 1 protection.

Algorithm 5: Generate safe groups

```
while |V| > 0 do
    u^s = the node with maximum degree in V;
    V = V − u^s;
    group g = new group {u^s};
    C = C ∪ {g};
    while |g| < k do
        Set candidates = { };
        for Each node u in V do
            if Adding u into g does not violate SGC then
                candidates = candidates ∪ {u};
        if |candidates| > 0 then
            for Each node u in candidates do
                u.cost = the cost to add u into g;
            u′ = u in candidates with minimum u.cost;
            g = g ∪ {u′}; V = V − {u′};
        else
            break;
    for each g in C with |g| < k do
        C = C − {g};
        for each node u in g do
            g′ = the group in C with the min. cost to add u under SGC;
            g′ = g′ ∪ {u};
```

4.4 Level 2 Protection

At Level 1, the *label-node* mapping is hidden by generalization. However, if an attacker knows both the label and the degree information of some nodes, he can still successfully re-identify some nodes in G_A. For example, if an attacker knows Bob's degree is 3, he can immediately find that node d is Bob in Fig. 3b. Therefore, in order to achieve the three protection objectives, the node re-identification using node label and degree information should be avoided for the nodes in set_{l_2}.

We generate a new published graph based on G_{L_1}. The target is to make sure that for each node in set_{l_2}, there are at least another *k-1* nodes in G_A identical to this node with respect to the attacker's background knowledge. At Level 2, we assume that an attacker also uses degree information, thus, the k-degree anonymous model [13] is needed to prevent the attack. k-degree anonymous model guarantees that the degree of each node in set_{l_2} appears at least k times in the published graph. To implement k-degree anonymous model based on G_{L_1}, for any node u in set_{l_2}, there should exist k nodes with the same degree as u and having u's label list. In this section, we call a graph that satisfies the above property ***k-degree anonymous graph***. We can construct a k-degree anonymous graph by making the nodes in the same group have the same degree if that group contains nodes in set_{l_2}.

In order to guarantee the privacy objectives (2) and (3), during the process of modifying G_{L_1} to a k-degree anonymous graph, *SGC* should be maintained. The k-degree anonymous graph generation method introduced in [13] does not satisfy the safe grouping constraint, thus we design a noise node/edge adding algorithm to construct the k-degree anonymous graph under *SGC*. Although the privacy protection problem in this section is different from Sect. 3, the adding noise node idea can still be adopted to construct the published graphs. The algorithm contains two steps:

- Generate a k-degree anonymous graph by adding nodes/edges under *SGC*;
- Since the new created nodes/edges do not have labels, assign labels to them.

We call the newly created nodes/edges "noise nodes/edges", next we discuss the detailed steps of adding noise nodes/edges.

4.4.1 Adding Nodes/Edges into G_{L_1}

To construct a k-degree anonymous graph, we first make the nodes in the same group have the same degree if that group contains nodes in set_{l_2} by adding nodes/edges. Specifically, we use the following three steps to perform the node/edge adding.

The first step is to set a target degree for each node. For a node u, if the group g that u belongs to contains at least one node in set_{l_2}, we set the target degree to be the largest degree in g. Otherwise, we set u's target degree as its original degree. For a group g that contains at least one node in set_{l_2}, since all the nodes in g need to have the same target degree, we also call this target degree g's target degree ($g.target$).

After each node has a target degree, the second step is to randomly create edges between the nodes which need to increase their degrees under *SGC*. If all the nodes reach their target degrees, we get a k-degree anonymous graph and finish the construction. Otherwise, we continue to construct the graph by adding some noise nodes, which is the next step.

In the third step, we add noise nodes to enable all the nodes in G_{L_1} have their target degrees by connecting them with noise nodes under *SGC*. When we add noise nodes, these noise nodes should be involved in certain groups to make them indistinguishable from the original nodes in G_{L_1}. In order to achieve this, we can let the noise nodes form groups with size k themselves, meanwhile, we need to hide the noise nodes by making their degree to a pre-selected value $degree_{target}$. $degree_{target}$ is the degree that the maximum number of groups of size k in G_{L_1} have.[5]

[5]If all the noise nodes have degree $degree_{target}$, they are hidden into those groups who have the maximum number of original groups with the same degree. This makes the noise nodes mixed with as many original nodes as possible to avoid the filtering.

Next, we first introduce Algorithm 6 which can make all the nodes in G_{L_1} reach their target degrees using several groups of same degree noise nodes. Then we prove an important property of this algorithm about the number of added noise nodes and the target degrees. Based on this property, we finally show how to set input parameters of Algorithm 6 to make all noise nodes exactly have degree $degree_{target}$ without violating *SGC*. In this part, we'll use two constants: $Change_{total}$ and $Change_{max}$. $Change_{total}$ represents the sum of degree needs to be increased for all the nodes in G_{L_1}. $Change_{max}$ stands for the maximum sum of degree needs to be increased in each group of G_{L_1}.

Algorithm 6 has four inputs: graph G_{L_1}, groups C, privacy parameter k and X, where X is the number of noise node groups that will be used to construct the new graph. In Algorithm 6, we create X groups of noise nodes and arrange them as an $X \times k$ array A. Each row of this array is a group of noise nodes. We use $A[i][j]$ to represent the noise node at position $[i, j]$ in A. New edges are created between the noise nodes in A and the nodes whose degrees need to be increased in G_{L_1}. Each time, we let all the nodes in one group reach their target degrees by creating edges from the nodes in this group to the noise nodes in A in a column by column style (line 4–12). Column by column means if an edge is connected with the noise node $A[i][j]$, then the next edge created will connect to $A[i+1][j]$. If $A[i+1][j]$ does not exist ($i+1 == X$), the edge is connected to the first noise node in the next column ($A[0][(j+1)\%k]$). The first created edge is linked to $A[0][0]$. After we complete these steps, all the nodes in G_{L_1} reach their target degrees. It is easy to see all the noise nodes either have the degree $d = \left\lfloor \frac{Change_{total}}{X \times k} \right\rfloor + 1$ or $(d-1)$ now. To let all the noise nodes have the same degree d, next we use the following method to build connections between noise nodes: (1) Select two rows with the maximum number of $(d-1)$ degree nodes in A; (2) Randomly select two nodes with the degree $(d-1)$ in each row and create an edge between them; (3) Repeat the above two steps until all the rows only contain nodes with degree d.

Theorem 5. *Algorithm 6 constructs a k-degree anonymous graph without violating* SGC *and all the noise nodes have the degree* $degree_{target}$, *if*

$$degree_{target} = \left\lfloor \frac{Change_{total}}{X \times k} \right\rfloor + 1 \tag{1}$$

$$X \geq max(\left\lceil \frac{Change_{max}}{k} \right\rceil, 2) \tag{2}$$

$$(degree_{target} \times k \times X + Change_{total})\%2 \equiv 0 \tag{3}$$

The detailed proof can be found in [29].

After the second step of the node/edge adding, $Change_{total}$ and $Change_{max}$ are both constants. As we introduced at the beginning of the third step, $degree_{target}$ is selected as the degree that the maximum number of groups of size k in G_{L_1}

Algorithm 6: Degree anonymizing algorithm with X

```
Create X × k new nodes and store them in a X × k array A;
Set each line of nodes in A as a new group;
int i = 0, j = 0;
while ¬(each node in G_{L1} reaches its target degree) do
    Random select a g ∈ C that ∃ u ∈ g, u.degree < g.target;
    while ¬(each node in g reaches its target degree) do
        Random select a u ∈ g with u.degree < g.target;
        Create an edge between u and A[i][j];
        i = i + 1;
        if i == X then
            i = 0;
            j = (j + 1)%k;
d = ⌊Change_total / (X×k)⌋ + 1;
while A contains node with degree d − 1 do
    Random select two rows l1 and l2 in A with maximum number of d − 1 degree nodes;
    Random select u1 ∈ l1, u2 ∈ l2 that u1.degree = u2.degree = d − 1;
    Create an edge between u1 and u2;
```

have, thus, $degree_{target}$ is also a constant. By solving the formulas in Theorem 5, we can derive X for Algorithm 6. If an integer X does not exist for Eq. (1) or the Eq. (2)/(3) cannot be satisfied, we can get a proper X by increasing the value of $Change_{total}$. There are several methods to increase the value of $Change_{total}$. Increasing the target degree of a group that contains the odd number of nodes by 1 can increase $Change_{total}$ by an odd number.[6] Increasing the target degree of a group that contains the even number of nodes by 1 or a group that contains the odd number of nodes by 2 can increase $Change_{total}$ by an even number. Delete one noise edge added in the previous step can also increase $Change_{total}$ by 2.

After completing the three steps, for each node u in G_{L_1} that needs to increase its degree by x, we connect u with x nodes which belong to either G_{L_1} or newly added noise nodes. All the noise nodes connected with u are in different groups. Since the degree of the noise node groups is the same as the degree of some existing groups, the noise nodes are hidden into those original nodes.

4.4.2 Label Assignment

After adding the noise nodes/edges, we should assign labels to them. Suppose $No_{con}(A_1, A_2, l))$ is the number of edges in the original graph that connect label A_1 with A_2 through an edge with label l_x, for all A_1, A_2 and l, a node/edge label

[6] We can force the grouping algorithm to generate at least one group with the odd number of nodes that contains nodes in set_{l_2}.

assignment with less $No_{con}(A_1, A_2, l)$ changes is preferred. So we use the following heuristic method to do the assignment.

We first assign labels to the noise edges. For the edges that connect with at least one noise node, we randomly assign labels to them following the edge label distribution in the original graph. For an edge between two nodes u_1 and u_2 in G_{L_1}, denote u_1's label list as $(A_{u_1,1}, A_{u_1,2}, \ldots, A_{u_1,t})$ and u_2's label list as $(A_{u_2,1}, A_{u_2,2}, \ldots, A_{u_2,t})$, we select the edge label l_x that satisfies $max(\sum_{j=1}^{t} No_{con}(A_{u_1,j}, A_{u_2,j}, l_x))$. In it, $No_{con}(A_{u_1,j}, A_{u_2,j}, l_x)$ is the number of edges in the original graph that connect label $A_{u_1,j}$ with $A_{u_2,j}$ through an edge with label l_x. Since the new created connections between labels appear the maximum number of times in the original graph, selecting l_x tends to change the connection between labels as less as possible.

After each noise edge obtains a label, we decide the labels of noise nodes according to the assigned edge labels and the original graph's connectivity. For a noise node u, suppose the labels on the edges adjacent to u are: $\{l_1, l_2, \ldots, l_d\}$, the selection rule of u's label list $(A_{u,1}, A_{u,2}, \ldots, A_{u,t})$ is $max(\sum_{i=1}^{d} \sum_{j=1}^{t} No_{con}(A_{u,j}, A_{u_{e_i},j}, l_i))$. In it u_{e_i} is the node connecting with u through edge i, $A_{u_{e_i},j}$ is u_{e_i}'s label on dimension j, and $No_{con}(A_{u,j}, A_{u_{e_i},j}, l_i)$ is the number of edges in the original graph that connect label $A_{u,j}$ with $A_{u_{e_i},j}$ through an edge with label l_i. We use the above formula to select the node labels which make the new created connections between labels appear the maximum number of times in the original graph. When all the noise nodes have labels, the graph can be published by generalizing the node labels of each noise node group, which is denoted as G_{L_2}.

4.5 Level 3 Protection

At Level 3, an attacker also knows the labels on the edges adjacent to set_{l_3}. Then, in this case, the anonymous graph generated at Level 2 cannot prevent the node from re-identification if the attacker also uses the edge label information. For example, in Fig. 3c, c and f are assigned to the same group and have the same degree. If an attacker knows that Tom is a 35 year-old man from Africa who has two neighbors both with type 1 (roommate), he can use this information to re-identify that c is Tom. In order to prevent such an attack, at level 3, when an attacker uses node label list, degree and edge label information of any node in set_{l_3} to search the published graph, the real matching node should be mixed with at least $k - 1$ other nodes. We call a graph that satisfies the above property ***k-degree-label anonymous graph***. We call the vector of labels on the edges adjacent to a node u u's ***degree label sequence***. For example, Fig. 3d is a 2-degree-label anonymous graph. The degree label sequence of node d is $[1, (0, 1), (0, 1)]$.

We can change a k-degree anonymous graph to be a k-degree-label anonymous graph by generalizing the labels on edges. That is, if a group contains at least one node that needs Level 3 protection, for any two nodes in it, change their degree label sequence to be the same by generalizing edge labels.

4.6 *Summary*

In this section, we design a comprehensive privacy protection framework for the labeled social networks. This framework allows different users to set different privacy protection requirements. This framework provides the possibility of offering different services to different users.

5 Protect You More Than Blank: Anti-learning Sensitive User Information in the Social Networks

5.1 *Motivation*

When a user registers an account on a social network (for example, Facebook), he is often required to fill in some personal information to create an online profile, including hobby, high school name, age, and so on. We call each input data item a *label* or an *attribute* of the user. For example, in Fig. 4 each user has two labels, the hobby and the purpose for joining the network. The weight between any two users represents the frequency they post on each other's wall. The frequency is normalized to a number in [0, 10].

Due to personal and cultural diversities, users have different concerns about sensitive labels. Some users may not mind others acquiring certain labels. Some may even want more people to know about them by publishing their labels. These are referred to as *open users* in the network. For example, in Fig. 4, Harry, Michelle, and so on, would like to share their hobbies with friends. However, some users treat this label as sensitive information, and do not want others to know. We call these users *conservative users*. For these conservative users, the websites normally leave their sensitive information blank to satisfy their privacy requirements. Ben in Fig. 4, for instance, is a very sensitive person and want to hide his own hobby information. As a result, we only obtained a graph as shown in Fig. 4, that is, Ben's hobby is left blank.

However, simply leaving sensitive labels blank is not enough to protect users' privacy. A group of learning algorithms have been introduced in [40, 48] to deduce the missing labels in the social networks with high accuracy. The attacks through links, groups and their combinations could detect the sensitive labels without any notice to users. These algorithms use the fact that users often have common social actions in a network, that is individuals who have similar labels tend to form a community. For example, although Ben in Fig. 4 does not want to publish his hobby, we could guess it as cooking since two friends of Ben (Harry and Bily) like cooking and only one friend (Aron) likes writing.[7] If Ben indeed like cooking, then his

[7] We can draw the same conclusion when considering the edge weights.

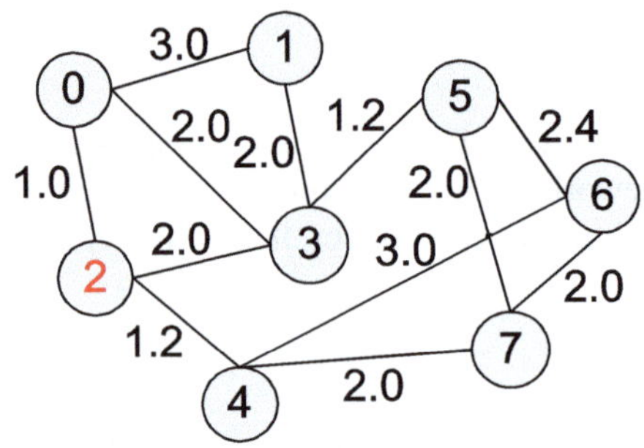

ID	NAME	HOBBY	PURPOSE
0	Harry	cooking	friendship
1	Michelle	cooking	relationship
2	Ben	*	friendship
3	Bily	cooking	networking
4	Aron	writing	dating
5	Sam	writing	relationship
6	San	writing	friendship
7	Tony	writing	dating

Fig. 4 Social network with blank label

privacy is violated even when the website protects him by leaving his hobby blank. Therefore, to avoid users' sensitive labels being learned through their neighborhood labels, we should break the association between his label and his neighborhood labels. In other words, we want to publish a social graph which does not disclose sensitive labels when learning algorithms are being applied. (It should be noted that the problem cannot simply be solved by replacing the blank attribute with a random label. When the associations between label and neighborhood are not broken, they can still be used to filter out the randomly filled labels.) Therefore, it is necessary to design a new model to protect sensitive information from being attacked by a learning algorithm and to meanwhile publish useful information for sharing.

In this section, we propose methods to break the label-neighborhood association based on user's personalized settings before releasing the social network. For open users, their real labels are published without change. For each conservative user, we modify his neighborhood and make the probability that an attacker can correctly guess his sensitive labels is at most p. p is published with the graph and attackers know the method we used to modify the graph. The modification is done by editing the edges. During the modification process, we first randomly select a set of labels S from other possible labels. Then we edit the edges to make sure the selected labels have more impact than the real labels based on learning algorithms. For example, Ben in Fig. 4 is a conservative user and $p = 50\,\%$. Through our method, we can change his neighborhood and make "writing" impact Ben more as shown in Fig. 5 (deleting the edge between 2 and 3). Since S is randomly selected, an attacker could not trust the learning result of any conservative node with confidence more than p.

In order to increase the utility of the published graph, we also preserve important characteristics of the social network, "the influential values" of nodes. As a common graph property, influential value is widely used in different applications, such as expert finding, the most valuable people searching, and so on [8, 49]. The value represents how important a user is in the whole network or to his neighbors. If each user has the same influential value in the published graph as in the original graph, he has the same "position" in the two graphs. Therefore, the influential value should be preserved in order to keep a graph as useful as before. We achieve the

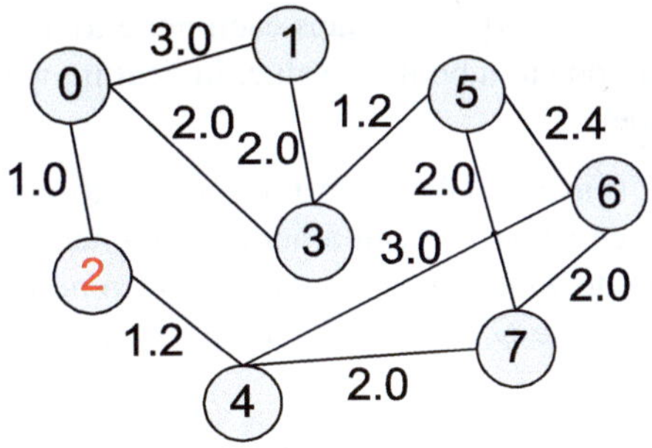

ID	NAME	HOBBY	PURPOSE
0	Harry	cooking	friendship
1	Michelle	cooking	relationship
2	Ben	*	friendship
3	Bily	cooking	networking
4	Aron	writing	dating
5	Sam	writing	relationship
6	San	writing	friendship
7	Tony	writing	dating

Fig. 5 Edge editing to change Ben's neighborhood

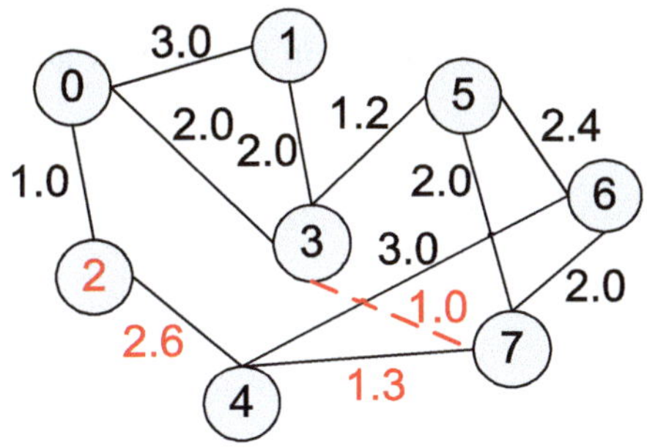

ID	NAME	HOBBY	PURPOSE
0	Harry	cooking	friendship
1	Michelle	cooking	relationship
2	Ben	*	friendship
3	Bily	cooking	networking
4	Aron	writing	dating
5	Sam	writing	relationship
6	San	writing	friendship
7	Tony	writing	dating

Fig. 6 Edge editing to reserve the influential values

influential value preserving objective by edge editing, too. An interesting result we observed is through preserving the influential values, the maximum eigenvalue of the graph's adjacent matrix is also preserved. Eigenvalues have been shown intimately connected to many important topological features in [26]. Figure 4 is an influential preserved graph of Fig. 4 if Ben wants to protect his label. If using the propagation based expert computation model in [8, 50] to analyze Figs. 4 and 6, the corresponding nodes in these two figures have the same influential value.

5.2 *Problem Description*

The social network is represented as an un-directed weighted graph.

Definition 6. Social network graph: a social network graph is a six tuple $G(V, E, W, \sigma, \sigma', \lambda)$, where V is a set of nodes, and each node represents an entity in the social network. $E \subseteq V \times V$ is the set of edges between nodes. $W : E \longrightarrow R^+$ maps each edge to a positive real number. σ maps each node to a group of non-sensitive labels. $\sigma' : V \longrightarrow s|s \in S$ maps each node to a sensitive label. $\lambda : V \longrightarrow boolean$ maps each node to a boolean value, indicating whether a node wants to protect its sensitive label.

We assume each node only has one sensitive label. If a node wants to protect its sensitive label, λ is set as true and vice versa. We call a node which needs to hide its label a conservative node and a node which does not care an open node. In the remaining part of this chapter, we also use G or $G(V, E, W)$ to represent a graph $G(V, E, W, \sigma, \sigma', \lambda)$ for simplicity.

For a weighted graph, one important data mining task is to analyze the node influential values (i.e. the expert finding [8, 49, 50]). We focus on the undirected weighted graph in which weights reflect the absolute cooperation between users. The node influential values in a weighted graph G are defined as:

Definition 7. The influential value f_i of each node u_i in a weighted graph G is defined as [8, 50]:

$$\forall u_i \in V, ft_i = f_i + \sum_j w_{(i,j)} \cdot f_j$$

$$T = \sum_{i=1}^{|V|} ft_i \tag{4}$$

$$\forall u_i \in V, f_i = \frac{ft_i}{T}$$

Equation (4) [8, 50] is a typical propagation model to compute node influential value. All f_is can be calculated when the above formulas converge at the stable state.

The *label-structure attack* is the attack which learns the blank label of a conservative node u by analyzing its neighborhood. The label l which has the most impact on u in u's neighborhood (l appears on at least one node in u's neighborhood) is most likely to be u's real label [40, 48, 51–53]. We call it *label-neighborhood relationship*. Specifically, we focus on the direct neighbors of each node, the attacks work as follows:

Definition 8. label-structure attack: the label l which has the most impact on u in u's neighborhood is set as u's real label. The impact value of l to u can be any of the following values:

- Frequency: the occurrence number of l in u's neighborhood.
- Weight: the total weight of nodes whose label is l in u's neighborhood: $\sum_{e(u,y) \wedge y.label=l} w_e$
- Influence: the related influential values of nodes whose label is l in u's neighborhood: $\sum_{e(u,y) \wedge y.label=l} w_e \cdot f_y$

The problem we solve in this section is:

Problem 1. Given a graph $G(V, E, W)$ and a constant p, generate a new graph $G'(V', E', W')$ from G that satisfies:

- $V \equiv V'$;
- The probability that an attacker could correctly identify the sensitive label of a conservative node through "label-structure attack" is at most p.[8]
- The influential values of any node u in G and G' are the same.

5.3 Algorithms

We generate a privacy preserved graph by edge editing. The operations include two parts: adding/deleting edges and adjusting the edge weights. Adding/deleting edges can be considered a special adjustment to edge weights. For a graph G, all edges have positive weights. A non-existence edge can be seen with weight 0. Adding one edge is to adjust the edge weight on an edge from zero to a positive value. Deleting one edge is to adjust the edge weight to zero.

For each conservative node, we make the association between its label and neighborhood uncertain by edge editing. Then we continue editing several edges to preserve the influential values of all the nodes. Therefore, our algorithm includes two parts:

- Breaking the label-neighborhood association.
- Preserving the influential values.

5.3.1 Edge Editing for Association Breaking

For a conservative node u, a learning algorithm can dig out u's label correctly because the same label is prevalent in the neighborhood of u in most cases. We break this association by reducing the edge weights.

Suppose there are totally L different labels in the social network and the real label on u is l. There are b label types appearing in its neighbors. Our algorithm edits u's neighborhood so that the probability an attacker trusts the learning result is at most p even with a learning algorithm which can reach 100 % accuracy. When we change u's neighborhood, we reduce the impact of l to u. We could either retain l in u's neighborhood or remove it.

We say label l_1 is ranking higher than label l_2 in u's neighborhood if l_1's impact value is larger than l_2. If we keep l in u's neighborhood, we should make sure that l

[8] In this section, we use the same p for all conservative nodes to demonstrate our algorithms. Our algorithms can support different ps as well.

is not always ranked first. Our algorithm moves l from first to any possible position in those b labels in u's neighborhood. The rank of l could be 1 to b. The probability he selects any label appearing in the ranked list is $\frac{1}{b}$, which should be less than p. Thus if $b \geq \lceil \frac{1}{p} \rceil$, we retain l in u's neighborhood. We move l to at least the kth (k is randomly selected) position in the ranked list by giving ks other labels larger impact values than l.

On the other hand, if $b < \lceil \frac{1}{p} \rceil$, the number of labels in u's neighborhood is not enough to protect l. Then we remove l out of u's neighborhood. Suppose the number of labels that do not appear in u's neighborhood is m after the adjustment. If an attacker observes that the number of labels in u's neighborhood is less than $\lceil \frac{1}{p} \rceil$, he knows our algorithm hides the real label in the labels that do not appear in u's neighborhood. Thus m should satisfy $\frac{1}{m} \leq p$. The remaining types of labels in u's neighborhood after the adjusting must $\leq (L - \lceil \frac{1}{p} \rceil)$. If only removing l cannot guarantee this condition, we randomly remove more connections until the number of labels in u's neighborhood reaches $(L - \lceil \frac{1}{p} \rceil)$.

To summarize, for any conservative node u, our edge editing method does the following operations:

- If $b \geq \lceil \frac{1}{p} \rceil$, we set the adjustment target to be k $(1 \leq k \leq \lceil \frac{1}{p} \rceil)$. We change the weights on the edges adjacent to u until there are k other labels which have larger impact values than its real label in the neighborhood;
- If $b < \lceil \frac{1}{p} \rceil$, we remove all u's adjacent nodes that have the same sensitive label out of its neighborhood. If the remaining label types are bigger than $(L - \lceil \frac{1}{p} \rceil)$, we remove more edges until it reaches $(L - \lceil \frac{1}{p} \rceil)$.

We adjust edge weights adjacent to u to make sure that there are at least k other labels whose impact values are bigger than l. In order to change the graph as little as possible, we implement the adjustment with the minimum sum of edge weight changes.

If we set the total edge weight changes less than or equal to a constant c, this adjustment problem can be modeled as a Satisfiability Modulo Theories (SMT) problem. Given a constant c, we can use a SMT solver to find one adjustment solution with the total weight changes $\leq c$. We can model the edge weights' adjustment problem as a SMT problem to find an adjustment solution. Since the SMT problem support $\neg$ and $\wedge$, all the first-order predicates are supported. The SMT model is:

- Variables
  - The amount of reduced weight on each edge:
    $\forall e(u, y) \wedge$y.label$\equiv l$, $\Delta w_{(u,y)}$:real$[0, w_{(u,y)}]$

- The variable that represents each changed edge: $\forall e(u, y) \wedge y.label \equiv l, c_{(u,y)}$: integer [0,1] (0: the edge still exists; 1: the edge is deleted)
- The number of nodes connected u with label l: n_l
- $w_l = \sum_{\forall e(u,y), y.label \equiv l}(w_{(u,y)} - \Delta w_{(u,y)})$
- $wi_l = \sum_{\forall e(u,y), y.label \equiv l}(w_{(u,y)} - \Delta w_{(u,y)}) \cdot f_y$
- Whether label l_i's frequency based impact value is bigger than l, 1 means bigger:

$$\forall l_i, s1_{l_i} : \text{integer}[0, 1]$$

- Whether label l_i's weight based impact value is bigger than l, 1 means bigger:

$$\forall l_i, s2_{l_i} : \text{integer}[0, 1]$$

- Whether label l_i's influence based impact value is bigger than l, 1 means bigger:

$$\forall l_i, s3_{l_i} : \text{integer}[0, 1]$$

- Whether label l_i's impact value is bigger than l no matter based on frequency, weight or influence, 1 means bigger:

$$\forall l_i, s_{l_i} : \text{integer}[0, 1]$$

- Constants
  - The maximum sum of edge weight changes: c
  - The privacy parameter: k
  - Number of nodes connected u with label l_i: N_{l_i}
  - $w_{l_i} = \sum_{\forall e(u,y), y.label \equiv l_i} w_{(u,y)}$
  - $wi_{l_i} = \sum_{\forall e(u,y), y.label \equiv l_i} w_{(u,y)} \cdot f_y$
- Constraints
  - The value of $s1_{l_i}$ is determined by N_{l_i} and n_l: $\forall l_i, ((s1_{1_i} \equiv 1) \wedge (N_{l_i} > n_l)) \vee ((s1_{1_i} \equiv 0) \wedge (N_{l_i} \leq n_l))$
  - The value of $s2_{l_i}$ is determined by w_{l_i} and w_l: $\forall l_i, ((s2_{1_i} \equiv 1) \wedge (w_{l_i} \geq w_l)) \vee ((s2_{1_i} \equiv 0) \wedge (w_{l_i} < w_l))$
  - The value of $s3_{l_i}$ is determined by wi_{l_i} and wi_l: $\forall l_i, ((s3_{1_i} \equiv 1) \wedge (wi_{l_i} \geq wi_l)) \vee ((s3_{1_i} \equiv 0) \wedge (wi_{l_i} < wi_l))$
  - s_{l_i} is determined by $s1_{l_i}$, $s2_{l_i}$ and $s3_{l_i}$: $\forall l_i, ((s_{1_i} \equiv 1) \wedge (s1_{l_i} + s2_{l_i} + s3_{l_i} \equiv 3)) \vee ((s_{1_i} \equiv 0) \wedge (s1_{l_i} + s2_{l_i} + s3_{l_i} < 3))$
  - There are at least k labels which have a larger impact value than l:

$$\sum_{\forall l_i} s_{l_i} \geq k$$

– Set $c_{(u,y)}$'s value:
$\forall c_{(u,y)}, ((c_{(u,y)} \equiv 1) \wedge (\Delta w_{(u,y)} < w_{(u,y)})) \vee ((c_{(u,y)} \equiv 0) \wedge (\Delta w_{(u,y)} \equiv w_{(u,y)}))$
– Set n_l's value:

$$n_l = \sum c_{u,y}$$

– the total weight changes should be at most c:

$$c \geq \sum \Delta w_{(u,y)}$$

If c is too small to find an adjusting solution, the SMT solver returns "unsat". If we set $c_{max} = \sum_{\forall e(u,y), y.label \equiv l}(w_{(u,y)})$ and $c_{min} = 0$, we can do a binary search on c to find a solution. The original value of c_{max} is the total weight changes when deleting all the edges that connect a node with the same label as u. Then the optimal solution must have the total weight changes within the range of $[c_{min}, c_{max}]$. We set $c = \frac{c_{min}+c_{max}}{2}$ and use SMT solver to solve the corresponding problem with this c. If a solution is found, then the optimal solution must be in $[\frac{c_{min}+c_{max}}{2}, c_{max}]$. If the SMT solver returns "unsat", the optimal solution must be in $[c_{min}, \frac{c_{min}+c_{max}}{2}]$. We repeat the process until we get close enough to the optimal solution (We set the terminate condition as $\frac{c_{max}-c_{min}}{c_{min}} < 5\,\%$). The efficiency of SMT solver depends on the problem size and structure. Normally, modern SMT solvers can solve a problem with thousands of variables within 100 ms. Our problem contains tens or hundreds of variables, therefore, it can be solved very quickly. In our experiment, on average, each invocation of SMT solver takes 40 ms.

After we get the result from the SMT solver, we can set the new weights for the corresponding edges.

5.3.2 Edge Editing for Influential Value Preservation

After breaking the label-neighborhood relationship, some nodes' influential values are changed. We use the following method to preserve the influential values:

- After processing several conservative nodes to achieve privacy protection, we try to build connections between all nodes whose neighborhoods are changed. If connections with positive edge weights which preserve the influential values are found, we add connections among these changed nodes only;
- If no solution is found, we do the adjustment for each edge whose weight is changed.

We first try to add connections among all the nodes whose influential values are changed. We want to preserve the influential values through the connections between the changed nodes while all the other nodes are not involved in. When we reduce the weight on the edge $e(u, v)$, if we can maintain both ft_u, ft_v and T in Eq. (4) unchanged, the influential value computation formulas still retain at the converging state. The influential values of all nodes are not changed either. We can

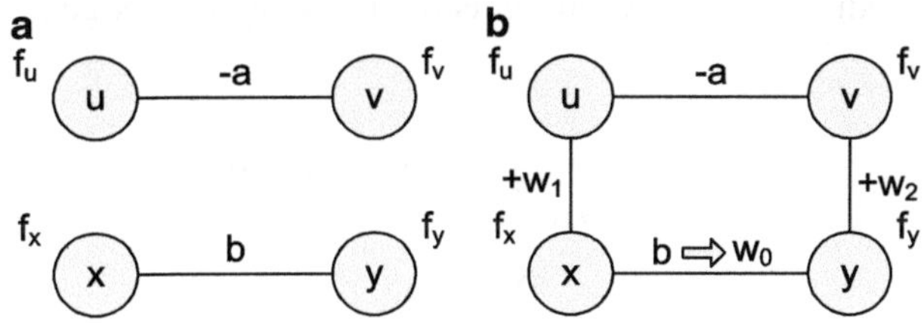

Fig. 7 Adjusting after deleting situation 1. (**a**) Original; (**b**) adjusting

model the connection creation problem as a group of linear formulas and use a SMT solver to find a solution. The modeling of this problem is similar to edge weights adjustment problem in Sect. 5.3.1.

For a sensitive node u, the change of its influential value is: $\sum_{e(u,y)} \Delta w_{(u,y)} \cdot f_y$,

Suppose the graph before the adjustment is $G_b(V_b, E_b, W_b)$, the weight on the edge (i, j) in G_b is $w_{(i,j)}$ and the edge weight after the adjustment is $w'_{(i,j)}$. The connection creation problem should satisfy the following constraints:

- The reduced influential value of each node is added back by the new connections:

$$\forall u, \sum_{e(u,y)} \Delta w_e \cdot f_y = \sum_{\forall j, j \neq u} (w'_{(i,j)} - w_{(i,j)}) \cdot f_j$$

- The connections just operated in the previous step should not be changed:

$$\forall (i, j) \in \{w_{i,j} \text{isjustchanged}\}, w'_{(i,j)} = w_{i,j}$$

- The weight on the edges must be positive numbers:

$$\forall (i, j), w'_{(i,j)} \geq 0$$

In it, $w'_{(i,j)}$ is the variable. For any two nodes i and j whose influential values are changed, we build a variable $w'_{(i,j)}$ for them. If there's no edge between i and j in G_b, $w_{(i,j)} = 0$.

If the above SMT problem does not have a solution, we use an adjustment method to add the changes of influential values back for each modified edge. Suppose an edge $e(u, v)$'s weight is reduced. We try to find another edge $e(x, y)$ that is the "closest" to u and v in the original graph. The "closest" means the average distance of $e(u, v)$ and $e(x, y)$: $\frac{dis(u,x)+dis(v,x)+dis(u,y)+dis(v,y)}{4}$ is the minimum among all edges. $e(x, u)$ should not be one of the edges changed while breaking the association. The influential value preservation is done through changing the weights on the links between x, y, u and v. We change the weights on the edges $e(u, x)$, $e(v, y)$ (if they do not exist, we first create these edges with weight 0) and adjust the weight on the edge $e(x, y)$. The operation is shown in Fig. 7. ft_u, ft_v, ft_x and ft_y should be preserved in order to retain f_u, f_v, f_x and f_y. Suppose the weight on the edge $e(u, v)$ is reduced by a. Before reducing the weight on edge e, the f_u, f_v, f_x and f_y satisfy:

$$ft_u = f_u + \sum_{\forall e(u,j) \wedge j \neq v} w_e \cdot f_j + a \cdot f_v$$

$$ft_v = f_v + \sum_{\forall e(v,j) \wedge j \neq u} w_e \cdot f_j + a \cdot f_u$$

$$ft_x = f_x + \sum_{\forall e(x,j) \wedge j \neq v} w_e \cdot f_j + b \cdot f_y$$

$$ft_y = f_y + \sum_{\forall e(y,j) \wedge j \neq v} w_e \cdot f_j + b \cdot f_x$$

After the reduction and adjustment operations, the f_u, f_v, f_x and f_y should satisfy:

$$ft_u = f_u + \sum_{\forall e(u,j) \wedge j \neq v} w_e \cdot f_j + w_1 \cdot f_x$$

$$ft_v = f_v + \sum_{\forall e(v,j) \wedge j \neq u} w_e \cdot f_j + w_2 \cdot f_y$$

$$ft_x = f_x + \sum_{\forall e(x,j) \wedge j \neq v} w_e \cdot f_j + w_1 \cdot f_u + w_0 \cdot f_y$$

$$ft_y = f_y + \sum_{\forall e(y,j) \wedge j \neq v} w_e \cdot f_j + w_2 \cdot f_v + w_0 \cdot f_x$$

To keep f_u, f_v, f_x, f_y, ft_u, ft_v, ft_x, ft_y and T unchanged:

$$a \cdot f_v = w_1 \cdot f_x$$

$$a \cdot f_u = w_2 \cdot f_y$$

$$b \cdot f_y = w_1 \cdot f_u + w_0 \cdot f_y$$

$$b \cdot f_x = w_2 \cdot f_v + w_0 \cdot f_x$$

We can get the new weights:

$$w_0 = \frac{b \cdot f_x \cdot f_y - a \cdot f_u \cdot f_v}{f_x \cdot f_y}$$

$$w_1 = \frac{a \cdot f_v}{f_x}$$

$$w_2 = \frac{a \cdot f_u}{f_y}$$

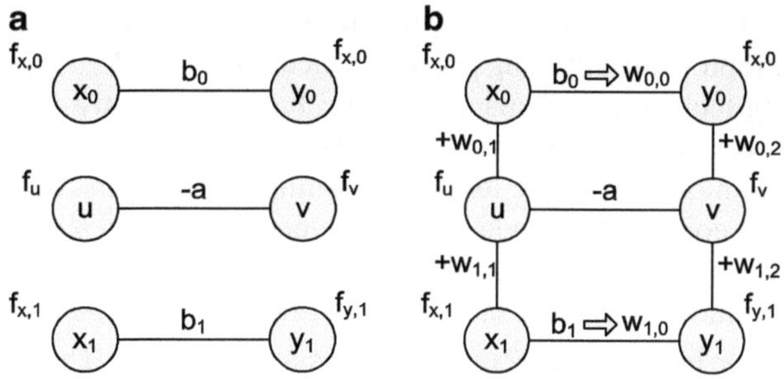

Fig. 8 Adjusting after deleting situation 2. (**a**) Original; (**b**) adjusting

We can see if there is an edge $e(x, y)$ with $b \cdot f_x \cdot f_y - a \cdot f_u \cdot f_v \geq 0$, we can always get a solution. For example, after deleting edge $e(2, 3)$ in Fig. 4, we select edge $e(4, 7)$ to keep the influential value unchanged.

In the case all $b \cdot f_x \cdot f_y - a \cdot f_u \cdot f_v < 0$, we use a group of edges to preserve the influential values. For each edge $e(x_j, y_j)$ in this group, we adjust the weights on edges $e(u, x_j)$, $e(v, y_j)$, and $e(x_j, y_j)$. Figure 8 shows an example of using two edges to do the adjustment.

When using a group of edges $e_(x_i, y_i)$ to make up the influential values, the new weights should satisfy:

$$a \cdot f_v = \sum_i w_{1,i} \cdot f_{x,i}$$

$$a \cdot f_u = \sum_i w_{2,i} \cdot f_{y,i}$$

$$\forall i, b_i \cdot f_{x,i} = w_{1,i} \cdot f_u + w_{0,i} \cdot f_{y,i}$$

$$\forall i, b_i \cdot f_{y,i} = w_{2,i} \cdot f_v + w_{0,i} \cdot f_{x,i}$$

One sufficient condition that the above formulas have a non-negative solution is: $\sum_i b_i \cdot f_{x,i} \cdot f_{y,i} \geq a \cdot f_u \cdot f_v$. In a large social network, there are enough edges to hold the above condition. The adjustment algorithm can always find a solution to make up the changed influential values.

5.4 Summary

In this section, motivated by users' different privacy preferences in social networks, we propose a new method to publish a graph for customized privacy protection. We found out that it is not secure after leaving users' sensitive information blank. Through some simple label-structure analysis, it is possible for attackers to know users' sensitive labels due to the label-structure association. Our method could provide protection against this attack. More importantly, we preserve one important metric of the graph while applying privacy protection.

The learning techniques in this section are basic data analysis methods which analyze the direct neighbors of each node (1-hop neighborhood). More advanced data mining techniques could be used for learning blank labels. Suppose a mining algorithm uses a conservative node u's x-hop neighborhood, a naive way to extend our approach is to ensure no node has the same label as u in u's x hop neighborhoods during the association breaking and influential preservation (the number of labels that do not appear in u's x hop neighborhood is at least $\lceil \frac{1}{p} \rceil$). This naive extension can guarantee the privacy protection objective. However, the utility of the published graph may be low since all the nodes with the same label as u will not appear in its x-hop neighborhood. In future works, we will consider better protection solutions for more complicated data analysis techniques.

6 A General Framework for Publishing Privacy Protected and Utility Preserved Graph

6.1 Motivation

The previous works [10, 13, 15, 16] prefer to publish a graph, which guarantees the protection of certain privacy with the smallest change to the original graph. However, there is no guarantee on how the utilities are preserved in the published graph. Only some empirical results are provided. While, preserving some utilities is the initial purpose to publish the graph. The utilities should be treated as important as privacy. Therefore, it becomes a challenge to quantify the utilities preserved in the published graph when we want to guarantee some privacy protection. Due to the conflicts between the privacy protecting and utility preserving, protecting privacy and preserving utilities normally cannot be achieved at the same time. It is preferred to have a general framework which can get a trade-off between the privacy and utility according to the data publisher' customized preferences.

In this work, we propose a fine-grained adjusting framework to publish a privacy protected and utility preserved graph (Fig. 9). There are three parameters: (1) The privacy requirements which specify the privacy to be protected; (2) The utility requirements which state the utilities to be preserved; (3) The publisher's preference on the trade-off. A "Conflict Extractor" is used to analyze the privacy and utility requirements to find the conflicting cores. These conflicting cores represent the conflictive points between privacy requirements and utility requirements. Then, an "Analyzer" is invoked to analyze the conflicting cores and find the best strategy, which provides the trade-off according to the publisher's preferences. The privacy/utility requirements are relaxed according to this trade-off strategy. We repeat the above process until a satisfiable result is computed.

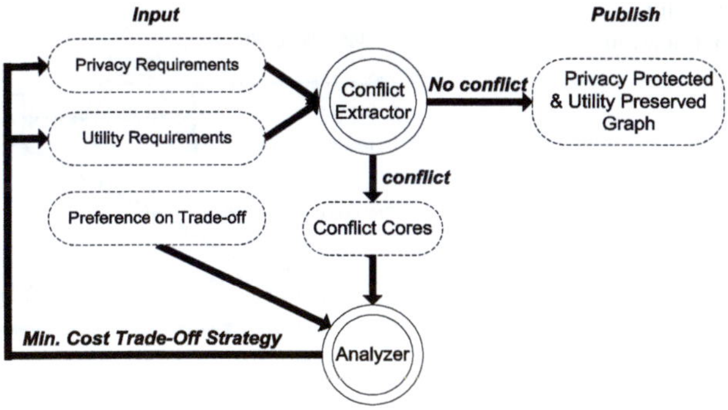

Fig. 9 Fine-grained data publishing framework

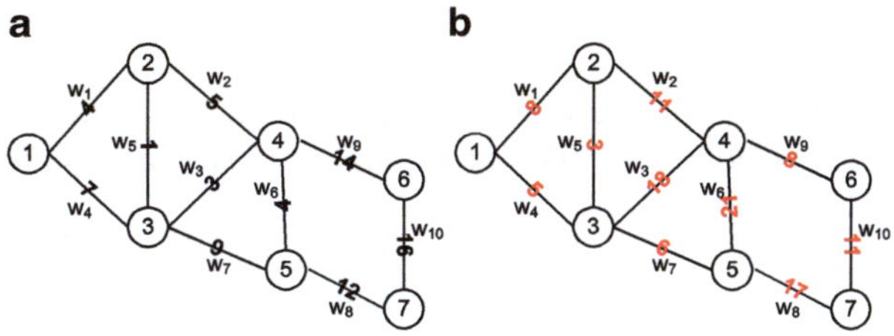

Fig. 10 Example Graph I. (**a**) Original; (**b**) published

We use the privacy preserving weighted graph publication problem [24, 25] as an example to demonstrate the implementation of our framework. The privacy preserving weighted graph publication problem aims at publishing a weighted graph in which the information about weights on edges are protected. Liu [24] and Das [25, 54] showed that many applications are modeled as weighted graphs. In these applications, the user connection information is public information and the weights on edges are the most important information to be protected. For a weighted graph, the magnitude of the weights as well as relative ordering of edge weights is the sensitive information to be protected [25, 54].

The basic method to protect edge weights is to set a new weight on each edge in the published graph. For example, Fig. 10b can be published for Fig. 10a. A lot of utility requirements in a weighted graph are linear inequations among edge weights [25]. We consider more general forms: the utility requirements are the graph properties that can be represented as linear inequations among edge weights connected by any first-order logic operations ($\wedge$, $\vee$, $\neg$ and any other derived operations such as XOR and $\Rightarrow$ etc.). We protect both the magnitude of edge weights as well as the relative ordering of some edges. The data publisher can freely specify the ordering between certain edge weights should be protected. For example, he may want to protect $ordering(w_1, w_4)$ and $ordering(w_2, w_3)$, which are part of the privacy requirements. The privacy requirements and utility requirements may not be

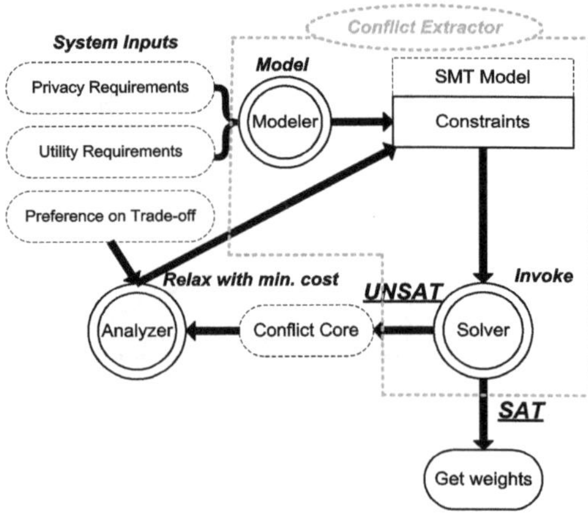

Fig. 11 Framework for weighted graph publication

satisfied at the same time since a utility or the combination of utilities may depend on some orderings of edge weights. For example, considering the following three utilities:

$$u_1 \ \ \mathit{Length}(1 \to 2 \to 4) < \mathit{Length}(1 \to 3 \to 4)$$
$$u_2 \ \ \mathit{Length}(2 \to 3 \to 4) < \mathit{Length}(2 \to 4)$$
$$u_3 \ \ \mathit{Length}(2 \to 4 \to 5) < \mathit{Length}(2 \to 3 \to 5) \tag{5}$$

They can be represented as: $u_1(w_1 + w_2 < w_3 + w_4)$, $u_2(w_3 + w_5 < w_2)$ and $u_3(w_2 + w_6 < w_5 + w_7)$.

The utility u_2 depends on the ordering $w_3 < w_2$ (If u_2 is preserved, w_3 must be smaller than w_2). u_2 and $\mathit{ordering}(w_2, w_3)$ are conflicting. The combinations of utilities u_1 and u_2 depend on the ordering $w_1 < w_4$. u_1 and u_2 are conflicting with $\mathit{ordering}(w_1, w_4)$. To guarantee the privacy protection, we must relax the utility requirements to make sure that they do not restrict any specific orderings between edge weights to be protected. After doing this, we could use a solver with a random seed to assign new edge weights based on the relaxed utility requirements. Since the relaxed utility requirements do not depend on the actual magnitude of edge weights or the ordering information of edges which the publisher wants to protect, the magnitude and ordering are protected. Of course, we can also ignore the protecting of some orderings to preserve more utilities.

We implement our framework for the privacy preserving weighted graph publication by taking the advantage of the Satisfiability Modulo Theory (SMT) (shown in Fig. 11). SMT problem is a decision problem for logical formulas, which are combinations of certain background theories (e.g. linear arithmetic) expressed in first-order logic. A SMT solver checks whether a SMT problem is satisfiable and returns the "evidence" of the result. When a SMT problem is satisfiable, the solver assigns values to the variables in the model. When a SMT problem is unsatisfiable, a conflicting core is returned to represent the reason why this problem is unsatisfiable. We model the utility/privacy requirements as SMT problems and use the SMT's conflicting core extraction to find the conflicting cores between privacy and utility. Then the "Analyzer" is designed to decide which requirements in conflicting cores should be removed.

6.2 Analysis of Previous Works

All the previous works did not consider how to do the fine-grained adjustment between privacy and utility. There is no guarantee on how utilities can be preserved in the published graph. Li [55] considered the trade-off between privacy and utility for tabular data publication. They get the trade-off by selecting one model from different protection models instead of relaxing privacy/utility requirements. Moreover, they cannot guarantee the preserving of utilities. Ashwin [56] considered the privacy and utility trade-off in social network recommendation, which is different from the graph publication problem.

6.3 Works on Edge Weight Protection

Besides the protection of node/link, two works [24, 25, 54] considered the edge weight protection when publishing a weighted graph. The basic method to protect edge weights is setting new weights on edges [24, 25, 54]. Liu [24] proposed a gaussian randomization multiplication strategy to set new weights. The edge weights are changed by adding noises according to a gaussian distribution. By doing this, a certain portion of shortest path lengths is preserved. This method only protects the magnitude of edge weights. The orderings of edges are not considered. Furthermore, only the length instead of the shortest path itself is considered.

Das [25, 54] proposed a Linear Programming (LP) based method, which preserves certain graph properties such as the shortest path etc. They model certain graph properties as linear inequalities and use a LP solver to generate new weights. For example, to keep the shortest path from node 1 to node 3 in Fig. 12 unchanged, a group of linear constraints as shown in Eq. (6) can be built [25, 54].

$$\begin{array}{ll} w_2 < w_1 & w_2 < w_3 \\ w_2 < w_4 & w_2 + w_5 < w_1 \\ w_2 + w_6 < w_3 & w_2 + w_5 < w_4 \\ w_2 + w_5 < w_2 + w_6 & w_2 + w_5 + w_7 < w_2 + w_6 \\ w_2 + w_5 + w_7 < w_4 & w_2 + w_5 + w_7 + w_8 < w_4 \end{array} \tag{6}$$

Fig. 12 Example Graph II

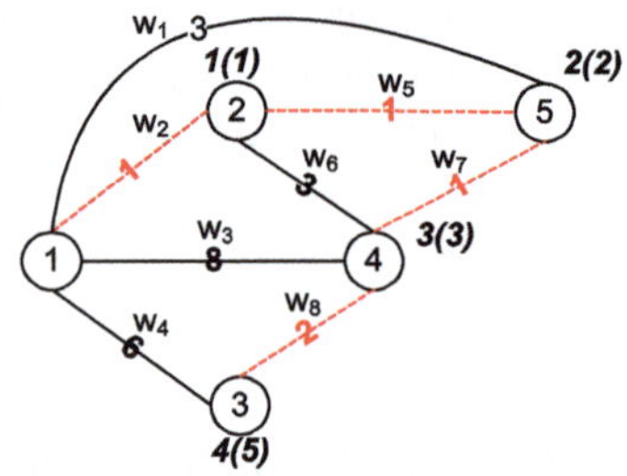

This method helps to keep the utilities in the published graph. Since the constraints are not based on the actual magnitudes of edge weights, the edge weight magnitude are protected when the values are randomly set by a solver [25]. However, the constraints arc built on graph properties without considering the protection of edge weight orderings. At the end of their paper [54], they mentioned that, to protect edge weight orderings, the k-anonymity of edge weights can be forced to be satisfied by setting all the edge weights in each node's edge neighborhood indistinguishable. Besides the overmuch implementation of k-anonymity, there are still two problems:

- The privacy protection may not be guaranteed because of the utility requirements themselves. For example, from the constraints in Eq. (6), we can find $w_{1,2} < w_{1,5}$, $w_{1,2} < w_{1,4}$ and $w_{1,2} < w_{1,3}$. It is obvious that edge $(1,2)$ cannot be anonymized with any $(1, v_i)$. While if we force $w_{1,2}$ equal to some other w_{1,v_i}s, LP solver cannot find any solution.
- The k-anonymity of edge weights cannot guarantee the protection of edge weight orderings. For example, if in the published graph, $(w_{1,5} = w_{1,2}) < (w_{1,4} = w_{1,3})$ (the edges next to node 1 satisfy 2-anonymity), the fact such that $w_{1,5} < w_{1,4}$ and $w_{1,2} < w_{1,4}$ etc., are still released.

Our framework protects both the edge weight magnitudes and orderings, which does not have the above shortcomings. We protect orderings by removing the dependency between the preserved utilities and protected orderings.

6.4 Framework and Case Study

6.4.1 General Framework

We propose a fine-grained adjusting framework (Fig. 9) to publish the privacy protected and utility preserved graph. The algorithm for this framework is shown in Algorithm 7. Besides the privacy requirements Pry and utility requirements *Utl*, the publisher can set his customized preference t on the trade-off. Whenever the published data G', which satisfies both Pry and Utl, cannot be found, we do relaxation on Pry and Utl according to t. We use a "Conflict Extractor" to analyze the privacy and utility requirements. It extracts the conflicting cores C between privacy and utility requirements. Then, we use an "Analyzer" to study C and find the best strategy sty which can get the trade-off according to t. We relax the corresponding privacy/utility requirements using sty and repeat the above process until G' can be generated. Finally, the framework generates G' which satisfies the relaxed privacy/utility requirements. The challenges to implement Algorithm 7 are: (1) How to construct "Conflict Extractor" to get the conflicting cores; (2) How to construct "Analyzer" to generate the adjusting strategy on the conflicting cores. Next, we demonstrate how to address these two challenges.

6.4.2 Privacy Preserving Weighted Graph Publication

The Privacy Preserving Weighted Graph Publication is to publish a graph with new edge weights under two requirements [25, 54]: Utility Requirements (the graph properties modeled as linear inequations connected by any first-order logic operations) and Privacy Requirements (magnitudes and Relative ordering between edge weights[9]).

Algorithm 7: Adjusting framework

while ¬(*G' satisfying both Pry and Utl can be constructed from G*) **do**
  Conflicting cores C = Conflit_Extractor(Pry, Utl);
  Strategy sty = Analyzer(C, t);
  Pry = relaxed Pry follows sty;
  Utl = relaxed Utl follows sty;
Generate G' from G which satisfies both Pry and Utl;
return G';

When using a solver to assign new edge weights based on the utility requirements, the linear constraints are not based on the actual magnitude of edge weights [25]. Therefore, the magnitudes of the edge weights are protected [25]. Similarly, to protect the orderings, given the utility requirement U_{req}, for any two edge weights w_x and w_y which the publisher does not want to release the ordering, we must guarantee the Free Ordering Rule defined as follows:

[9]We use the ordering between two edge weights to demonstrate our framework. Each privacy requirement can be directly replaced by a general form such as any linear inequations connected by first-logic operations without any change.

Definition 9. Free Ordering Rule (Free(U_{req}, p)): the utility requirements U_{req} satisfy Free Ordering Rule on the edge weight pair $p(w_x, w_y)$ if $\neg(U_{req} \Rightarrow (w_x < w_y)) \wedge \neg(U_{req} \Rightarrow (w_x > w_y))$.

Then the problem solved in this section is:

Problem 2. Privacy preserving weighted graph publication:

- Input:
  - A weighted graph G
  - A group of utility requirements where each utility u_i has a weight $u_i.w$ to represent its importance: $Utl = \{u_1(u_1.w), \ldots, u_m(u_m.w)\}$
  - A group of edge weight pairs where each pair p_i has a weight $p_i.w$ to represent its importance: $Pry = \{p_1(w_{1,a}, w_{1,b})(p_1.w), \ldots, p_n(w_{n,a}, w_{n,b})(p_n.w)\}$
  - Thc privacy objective: threshold t, which represents the publisher's preference on trade-off.
- Output: A weighted graph G' with new edge weights: $Utl_{G'} = \{u_i | \forall u_i \in Utl \wedge u_i \in G'\}$ and the edge weights in G' are assigned by a solver only based on $Utl_{G'}$, where:

$$minimize: \sum_{\forall u_i \in Utl \wedge u_i \notin G'} u_i.w$$

$$with:$$

$$\sum_{\forall p_i \in Pry \wedge Free(Utl_{G'}, p_i)} p_i.w \geq t$$

The input of the problem includes a weighted graph, the utility/privacy requirements and a threshold t. Each utility requirement has a weight which represents how important the utility is. The privacy requirements contain edge weight pairs with importance. The threshold t is the privacy objective, which represents the publisher's preference on trade-off between the privacy and utility. Larger t is preferred for higher privacy and smaller t is preferred for higher utility. The publisher can freely set the utility/privacy requirements and their importance. For a utility which contains many inequations such as Eq. (6), the publisher can separate them into different parts and assign importance for these parts. The output graph G' guarantees the sum of importance on the edge pairs whose orderings are protected is at least t with the minimum sum of utility loss. For a utility u_i, if the published graph G' preserves u_i, we represent it as $u_i \in G'$. Then, the utilities still preserved in G' are $Utl_{G'} = \{u_i | \forall u_i \in Utl \wedge u_i \in G'\}$. The utility loss we want to minimize is $\sum_{\forall u_i \in Utl \wedge u_i \notin G'} u_i.w$ where $\sum_{\forall p_i \in Pry \wedge Free(Utl_{G'}, p_i)} p_i.w \geq t$. After getting $Utl_{G'}$, we use a solver to randomly (use the random seed parameter in a SMT solver) assign edge weights only based on $Utl_{G'}$.

6.4.3 Framework for Weighted Graph Publication

Adjusting Framework (Fig. 11):

- Step 1: Firstly, we model the utility and privacy requirements as SMT problems and use a SMT solver to find the conflicting cores. We put these conflicting cores together as a summarized conflicting core C.
- Step 2: If the privacy threshold t is reached, we go to next step. Otherwise, we analyze the summarized conflicting core to find the strategy to relax the utility requirements which minimize the utility loss when privacy protection is guaranteed to reach the threshold t. Then, we go back to step 1.
- Step 3: We build a SMT model on the current relaxed utility requirements and use a SMT solver to assign new edge weights for the published graph.

The framework is implemented by Algorithm 8. When the summation of importance on the pairs (privacy requirements) which do not appear in the summarized conflicting core C reaches the threshold t, the privacy objective is achieved (We represent a pair p_i appearing in C as $p_i \in C$ and a utility u_i appearing in C as $u_i \in C$.).

Algorithm 8: Adjusting framework for Problem 2

$pry_{gain} = 0$; $Utl_{G'} = Utl$; $P_{remain} = Pry$;
while *true* **do**
    Build SMT Models on P_{remain} and $Utl_{G'}$; `/* Step 1 */`
    Invoke a SMT solver to get the summarized conflicting core C;
    $pry_{gain} = \sum_{\forall p_i \notin C} p_i.w$; `/* Step 2 */`
    if $pry_{gain} < t$ **then**
        Find the removing utility set Utl_{rm} (Utl_{rm}=Algorithm 9(C));
        $Utl_{G'} = Utl_{G'} - Utl_{rm}$;
        $P_{remain} = P_{remain} - \{p_i | p_i \notin C\}$;
    else
        break;
Build a SMT model on the relaxed utilities $Utl_{G'}$; `/* Step 3 */`
Use a SMT solver to find new edge weights and generate G';
return G';

Conflict Extractor: For a pair of edge weights $p(w_x, w_y)$ whose ordering is preferred to be hidden by the publisher, without loss of the generality, we suppose the original ordering is $w_x > w_y$. Suppose U_c is the group of constraints for the utility requirements, we use a SMT solver to check whether $U_c \wedge w_x < w_y$ (when a privacy requirement has a general form P, we check $U_c \wedge \neg P$) is satisfiable ($U_c \wedge w_x > w_y$ must be satisfiable since U_c is extracted from the original graph). If $U_c \wedge w_x < w_y$ is unsatisfiable, a conflicting core will be returned. Finally, we can get a summarized conflicting core.

Analyzer: For an edge weight pair, if we remove any utility from its corresponding conflicting core, this conflicting core is disabled. We use $core_i$ to represent the conflicting core of p_i and $u_j \in core_i$ to represent the fact that u_j appears in $core_i$. For an edge pair $p_i \notin C$, its ordering has been protected since the utility requirements do not limit any special ordering. Thus, we should make the summation of importance on the pairs in C whose conflicting cores being disabled at least $t - \sum_{\forall p_i \notin C} p_i.w$. To find the best adjusting strategy according to the current summarized conflicting core, we should solve:

Problem 3. Finding the utility set Utl_{rm} to remove:

$$minimize: \sum_{\forall u_i \in Utl_{rm}} u_i.w$$

$$with:$$

$$\sum_{\forall p_i \in C \wedge (\exists u_j (u_j \in core_i \wedge u_j \in Utl_{rm}))} p_i.w \geq t - \sum_{\forall p_i \notin C} p_i.w$$

Theorem 6. *Problem 3 is a NP-Hard Problem.*

The Minimum Set-Cover problem to Problem 3 can be transformed to this problem.

Algorithm 9: Finding removing utilities

$Utl_{rm} = \{\}$ $\quad pry_{gain} = 0 \quad t' = t - \sum_{\forall p_i \notin C} p_i.w$;
while $pry_{gain} < t'$ **do**
  for *each* $u_i \in C$ **do**
    $u_i.score = \frac{u_i.w}{\sum_{\forall p_i \in C \wedge u_i \in core_i} p_i.w}$;
  $u_{min} = u_i$ with the minimum score where $u_i \in C$;
  $Utl_{rm} = Utl_{rm} \cup \{u_{min}\}$;
  $pry_{gain} = pry_{gain} + \sum_{\forall p_i \in C \wedge u_{min} \in core_i} p_i.w$;
  remove u_{min} and $\{core_i | \forall core_i \wedge (u_{min} \in core_i)\}$ out of C;

We use Algorithm 9 to compute the best removing utility set Utl_{rm}. pry_{gain} is a variable which records the sum of importance on the pairs whose conflicting cores are disabled by Utl_{rm}. We firstly compute the adjusting threshold of privacy $t' = t - \sum_{\forall p_i \notin C} p_i.w$. For each utility $u_i \in C$, we compute a score to represent the ratio between the loss of utility importance and the gain of privacy importance by removing u_i. We use the importance of u_i ($u_i.w$) dividing the summation of importance on the pairs whose cores are broken in C when removing u_i as u_i's score. Intuitively, deleting the utility with the minimum score will cause the minimum utility loss to gain the same privacy protection effect as deleting other utilities. So, we select the utility u_{min}, which has the minimum score in C, to delete. Deleting u_{min} disables all the cores which contain u_{min}. Then we update the privacy gain as

$pry_{gain} = pry_{gain} + \sum_{\forall p_i \in C \wedge u_{min} \in core_i} p_i.w$. We add u_{min} into Utl_{rm} and delete all the cores that contain u_{min}. If the privacy threshold is not reached ($pry_{gain} < t'$), we repeat the above process until $pry_{gain} \geq t'$.

It is obvious that if the threshold $t = 0$, a graph that satisfies all utility requirements is published. And if $t = \sum_{\forall p_i \in Pry} p_i.w$, a graph that satisfies all privacy requirements is published. Our framework could find the solutions on the trade-off curve between these two cases.

7 Summary

In this section, we proposed a fine-grained adjusting framework which can get the trade-off between privacy and utility. We used the privacy preserving weighted graph publication problem to demonstrate the implementation of our framework.

8 Conclusion

In this chapter, we discussed the privacy issues in social network data. The future works may include the following directions,

- Large scale graph publication
  The numbers of nodes in social networks such as facebook, linkedin and so on. are becoming extremely large nowadays. For example, the statistics by facebook shows there were 750 million active users by 2011 July. It is difficult and impossible to analyze a network with millions of nodes [57, 58]. Current graph protection models did not consider large scale graphs. A lot of protection models [14–16, 22] need to determine the isomorphism between subgraphs. While the graph isomorphism checking problem is neither known to be NP-complete nor to be tractable [59]. How to publish a privacy preserving large scale graph becomes an essential problem.
- Personal privacy protection in online social networks
  New challenges appear for privacy protection in online social networks. With the popularity of new communication methods such as Microblog, the information sharing becomes easier in social networks. This brings new challenges since a user may easily publish information about his friends.
- Privacy preserving graph publication in a distributed environment
  Another interesting direction is to consider how to implement the protection models in a distributed environment, where different publishers publish their data independently and their data are overlapping. In a distributed environment, although the data published by each publisher satisfy certain privacy requirements, an attacker can still intrude upon a user's privacy by combining the data

published by different publishers together [60]. Protocols should be designed to help these publishers publish unified data together to guarantee privacy.

- Background knowledge bounding by guiding user's privacy preference
  The main difficulty for privacy preserving graph publication is that it is hard to find a correct model to bound an adversary's background knowledge. If the assumption of the attacker's background knowledge is too strong, the utilities of the published graph are hard to preserve. If the assumption of the attacker's background knowledge is too weak, the privacy of the published graph is not preserved for all users. The privacy preference setting of each user in the online social network can be used to correctly bound the attacker's background knowledge. A standard model for an adversary's background knowledge can be built by modifying the privacy preference rules. How to do this to improve the quality of the published graph is an interesting future work.

References

1. Han J (2005) Data mining: concepts and techniques. Morgan Kaufmann, San Francisco
2. Backstrom L (2006) Group formation in large social networks: membership, growth, and evolution. In: In KDD 06: proceedings of the 12th ACM SIGKDD international conference on knowledge discovery and data mining. ACM Press, New York, pp 44–54
3. Baumes J, Goldberg M, Magdon-Ismail M, Wallace A (2004) Discovering hidden groups in communication networks. In: Proceedings of the 2nd NSF/NIJ symposium on intelligence and security informatics
4. Berger-Wolf TY, Saia J (2006) A framework for analysis of dynamic social networks. In: Proceedings of the 12th ACM SIGKDD international conference on Knowledge discovery and data mining (KDD '06), New York. ACM, New York, pp 523–528
5. Girvan M, Newman MEJ (2002) Community structure in social and biological networks. Proc Natl Acad Sci 99(12):7821–7826
6. Kempe D, Kleinberg J, Tardos E (2003) Maximizing the spread of influence through a social network. In: Proceedings of the ninth ACM SIGKDD international conference on Knowledge discovery and data mining (KDD '03), New York. ACM, New York, pp 137–146
7. Kumar R, Novak J, Tomkins A (2006) Structure and evolution of online social networks. In: Proceedings of the 12th ACM SIGKDD international conference on knowledge discovery and data mining (KDD '06), New York. ACM, New York, pp 611–617
8. Zhang J, Tang J, Li J (2007) Expert finding in a social network. In Proceedings of 12th International Conference on Database Systems for Advanced Applications, DASFAA, 1066–1069, 2007
9. Hay M, Liu K, Miklau G, Pei J, Terzi E (2011) Privacy-aware data management in information networks. In: Proceedings of the 2011 international conference on Management of data (SIGMOD '11), New York. ACM, New York, pp 1201–1204
10. Hay M, Miklau G, Jensen D, Towsley D, Weis P (2008) Resisting structural re-identification in anonymized social networks. Proc VLDB Endow 1(1):102–114
11. Zhou B, Pei J, Luk W (2008) A brief survey on anonymization techniques for privacy preserving publishing of social network data. SIGKDD Explor. Newsl. 10:12–22
12. Wu X, Ying X (2009) A survey of algorithms for privacy-preservation of graphs and social networks. Kluwer Academic, Dordrecht
13. Liu K, Terzi E (2008) Towards identity anonymization on graphs. In: SIGMOD '08: proceedings of the 2008 ACM SIGMOD international conference on Management of data, New York. ACM, New York, pp 93–106

14. Cheng J, Fu AW-C, Liu J (2010) K-isomorphism: privacy preserving network publication against structural attacks. In: proceedings of the 2010 ACM SIGMOD international conference on management of data (SIGMOD '10), New York. ACM, New York, pp 459–470
15. Zhou B, Pei J (2008) Preserving privacy in social networks against neighborhood attacks. In: ICDE '08: proceedings of the 2008 IEEE 24th international conference on data engineering, Washington, DC. IEEE Computer Society, Silver Spring, pp 506–515
16. Zou L, Chen L, Özsu MT (2009) k-automorphism: a general framework for privacy preserving network publication. Proc VLDB Endow 2(1):946–957
17. Ying X, Wu X (2008) Randomizing social networks: a spectrum preserving approach. In: Proceedings of the SIAM international conference on data mining, pp 739–750
18. Zheleva E, Getoor L (2008) Preserving the privacy of sensitive relationships in graph data. In: PinKDD'07: proceedings of the 1st ACM SIGKDD international conference on privacy, security, and trust in KDD. Springer, Berlin, pp 153–171
19. Campan A, Truta TM (2008) A clustering approach for data and structural anonymity in social networks. In: Privacy, security, and trust in KDD workshop (PinKDD)
20. Cormode G, Srivastava D, Yu T, Zhang Q (2008) Anonymizing bipartite graph data using safe groupings. Proc VLDB Endow 1(1):833–844
21. Bhagat S, Cormode G, Krishnamurthy B, Srivastava D (2009) Class-based graph anonymization for social network data. Proc VLDB Endow 2(1):766–777
22. Zhou B, Pei J (June 2010) The k-anonymity and l-diversity approaches for privacy preservation in social networks against neighborhood attacks. Knowl Inf Syst 28(1):47–77
23. Ford R, Truta TM, Campan A (2009) P-sensitive k-anonymity for social networks. In: DMIN 2009: proceedings of the 2009 international conference on data mining, pp 403–409
24. Liu L, Wang J, Liu J, Zhang J (2008) Privacy preserving in social networks against sensitive edge disclosure. Technical Report CMIDA-HiPSCCS 006-08
25. Das S, Egecioglu O, Abbadi AE (2011) Anonymizing weighted social network graphs. In: ICDE'11: proceedings of the 2011 IEEE 23th international conference on data engineering
26. Ying X, Wu X (2008) Randomizing social networks: a spectrum preserving approach. In: SDM. SIAM, Philadelphia, pp 739–750
27. Ying X, Wu X, Barbara D (2011) Spectrum based fraud detection in social networks. In: Proceedings of the 2011 IEEE 27th international conference on data engineering (ICDE '11), Washington, DC. IEEE Computer Society, Silver Spring, pp 912–923
28. Yuan M, Chen L, Yu PS, Yu T, Mei H (2011) Protecting sensitive labels in social network data anonymization. In IEEE Transactions on Data and Knowledge Engineering (TKDE), 25(3), pp 633–647, 2013
29. Yuan M, Chen L, Yu PS (2010) Personalized privacy protection in social networks. Proc VLDB Endow 4:141–150
30. Yuan M, Chen L, Rao W, Mei H (2012) A general framework for publishing privacy protected and utility preserved graph. In: Proceedings of the 2012 IEEE 12th international conference on data mining (ICDM '12), Washington, DC. IEEE Computer Society, Silver Spring, pp 1182–1187
31. Sweeney L (2002) k-anonymity: a model for protecting privacy. Int J Uncertain Quantif Fuzziness and Knowledge-Based Systems 10(5):557–570, 202
32. Machanavajjhala A, Kifer D, Gehrke J, Venkitasubramaniam M (2007) ACM Trans Knowl Discov Data (TKDD), 1(1), 2007
33. Bonchi F, Gionis A, Tassa T (2011) Identity obfuscation in graphs through the information theoretic lens. In: Proceedings of the 2011 IEEE 27th international conference on data engineering (ICDE '11), Washington, DC. IEEE Computer Society, Silver Spring, pp 924–935
34. Xiao Q, Wang Z, Tan K-L (2011) Lora: link obfuscation by randomization in graphs. In: Proceedings of the 8th VLDB international conference on secure data management (SDM'11). Springer, Berlin, pp 33–51
35. Backstrom L, Dwork C, Kleinberg J (2007) Wherefore art thou r3579x?: anonymized social networks, hidden patterns, and structural steganography. In: WWW '07: proceedings of the 16th international conference on World wide web, New York. ACM, New York, pp 181–190

36. Shrivastava N, Majumder A, Rastogi R (2008) Mining (social) network graphs to detect random link attacks. In: ICDE '08: proceedings of the 2008 IEEE 24th international conference on data engineering, Washington, DC. IEEE Computer Society, Silver Spring, pp 486–495
37. Narayanan A, Shmatikov V (2009) De-anonymizing social networks. In: Proceedings of the 2009 30th IEEE symposium on security and privacy, Washington, DC. IEEE Computer Society, Silver Spring, pp 173–187
38. Liu K, Terzi E (2009) A framework for computing the privacy scores of users in online social networks. In: Proceedings of the 2009 ninth IEEE international conference on data mining (ICDM '09), Washington, DC. IEEE Computer Society, Silver Spring, pp 288–297
39. Becker J, Chen H (2009) Measuring privacy risk in online social networks. In: Proceedings of the 2009 Web 2.0 security and privacy (W2SP '09)
40. Zheleva E, Getoor L (2009) To join or not to join: the illusion of privacy in social networks with mixed public and private user profiles. In: Proceedings of the 18th international conference on World wide web (WWW '09), New York. ACM, New York, pp 531–540
41. Mislove A, Viswanath B, Gummadi KP, Druschel P (2010) You are who you know: inferring user profiles in online social networks. In: Proceedings of the third ACM international conference on Web search and data mining (WSDM '10), New York. ACM, New York, pp 251–260
42. Lindamood J, Heatherly R, Kantarcioglu M, Thuraisingham B (2009) Inferring private information using social network data. In: Proceedings of the 18th international conference on World wide web (WWW '09), New York. ACM, New York, pp 1145–1146
43. Fang L, LeFevre K (2010) Privacy wizards for social networking sites. In: Proceedings of the 19th international conference on World wide web (WWW '10), New York. ACM, New York, pp 351–360
44. Shehab M, Cheek G, Touati H, Squicciarini AC, Cheng P-C (2010) Learning based access control in online social networks. In: Proceedings of the 19th international conference on World wide web (WWW '10), New York. ACM, New York, pp 1179–1180
45. Barabási A-L, Albert R (1999) Emergence of scaling in random networks. Science 286:509–512
46. Puttaswamy KP, Sala A, Zhao BY (2009) Starclique: guaranteeing user privacy in social networks against intersection attacks. In: Proceedings of the 5th international conference on emerging networking experiments and technologies (CoNEXT '09), New York. ACM, New York, pp 157–168
47. Xiao X, Tao Y (2006) Anatomy: simple and effective privacy preservation. In: VLDB'06, pp 139–150
48. Mo M, King I (2010) Exploit of online social networks with community-based graph semi-supervised learning. In: Proceedings of the 17th international conference on neural information processing: theory and algorithms - Volume Part I (ICONIP'10). Springer, Berlin, pp 669–678
49. Campbell CS, Maglio PP, Cozzi A, Dom B (2003) Expertise identification using email communications. In: CIKM '03: proceedings of the twelfth international conference on information and knowledge management, New York. ACM, New York, pp 528–531
50. Zhang J, Tang J, Liu L, Li J (2008) A mixture model for expert finding. In: Proceedings of the 12th Pacific-Asia conference on advances in knowledge discovery and data mining (PAKDD'08). Springer, Berlin, pp 466–478
51. Lu Q, Getoor L (2003) Link-based classification. In: Proceedings of the international conference on machine learning (ICML), pp 496–503
52. Neville J, and Jensen D (2003) Collective classification with relational dependency networks. J Mach Learn Res 8:2007
53. Tang L, Liu H (2009) Relational learning via latent social dimensions. In: Proceedings of the 15th ACM SIGKDD international conference on knowledge discovery and data mining (KDD '09), New York. ACM, New York, pp 817–826
54. Das S, Egecioglu O, Abbadi AE (2010) Anonimos: an lp based approach for anonymizing weighted social network graphs. In: TKDE, pp pre-print

55. Li T, Li N (2009) On the tradeoff between privacy and utility in data publishing. In: Proceedings of the 15th ACM SIGKDD international conference on knowledge discovery and data mining (KDD '09), New York. ACM, New York, pp 517–526
56. Machanavajjhala A, Korolova A, Sarma AD (2011) Personalized social recommendations: accurate or private. Proc VLDB Endow 4(7):440–450
57. Leskovec J, Kleinberg J, Faloutsos C (2005) Graphs over time: densification laws, shrinking diameters and possible explanations. In: Proceedings of the eleventh ACM SIGKDD international conference on knowledge discovery in data mining (KDD '05), New York. ACM, New York, pp 177–187
58. Leskovec J, Faloutsos C (2006) Sampling from large graphs. In: Proceedings of the 12th ACM SIGKDD international conference on knowledge discovery and data mining (KDD '06), New York. ACM, New York, pp 631–636
59. Schöning U (1987) Graph isomorphism is in the low hierarchy. In: Proceedings of the 4th annual symposium on theoretical aspects of computer science (STACS '87), London. Springer, London, pp 114–124
60. Ganta SR, Kasiviswanathan SP, Smith A (2008) Composition attacks and auxiliary information in data privacy. In: Proceedings of the 14th ACM SIGKDD international conference on knowledge discovery and data mining (KDD '08), New York. ACM, New York, pp. 265–273

Social Media: The Evolution of e-Health Services

Tiziana Guzzo, Alessia D'Andrea, Fernando Ferri, and Patrizia Grifoni

Abstract The chapter analyses e-health services provided by different Social Media (collaborative projects, blogs, content communities, social networking sites, virtual games and virtual social worlds and video-chat) and introduces a Hybrid Cloud E-health Services architecture (HCLES) able to provide open, interoperable, scalable, and extensible services for all the e-health activities. It integrates the potentialities of Skype for a direct communication and synchronous data transmission among people with the cloud perspective of Social Media services.

1 Introduction

In the last years, the professional use of Social Media is growing more and more in different fields such as: business [1], learning [2], scientific research [3], tourism [4], etc. In health field advances in Social Media are changing the way healthcare services are provided and the term "e-health" is broadly used to describe this evolution. There is not a consensus on the definition of e-Health concept. Eysenbach [5] provides the following: "e-health is an emerging field in the intersection of medical informatics, public health and business, referring to health services and information delivered or enhanced through the Internet and related technologies". Marconi [6] defines e-health as "the application of Internet and other related technologies in the healthcare industry to improve the access, efficiency, effectiveness, and quality of clinical and business processes utilized by healthcare organizations, practitioners, patients, and consumers in an effort to improve the health status of patients". A similar definition is given in [7] "e-health is the use of Internet technology by the public, health workers, and others to access

T. Guzzo • A. D'Andrea (✉) • F. Ferri • P. Grifoni
IRPPS-CNR, via Palestro 32, 00185 Rome, Italy
e-mail: tiziana.guzzo@irpps.cnr.it; alessia.dandrea@irpps.cnr.it; fernando.ferri@irpps.cnr.it; patrizia.grifoni@irpps.cnr.it

Ş. Gündüz-Öğüdücü and A.Ş. Etaner-Uyar (eds.), *Social Networks: Analysis and Case Studies*, Lecture Notes in Social Networks, DOI 10.1007/978-3-7091-1797-2_10,

health and lifestyle information, services and support; it encompasses telemedicine, telecare, etc." In Akeh and Morfaw [8] e-health is defined as "the ability to use Internet technology to provide health services and deliver care to individuals from geographically dispersed locations".

Improving access to e-health has been receiving particular attention since the first World Telecommunication Development Conference (WTDC) in 1994. In 2005, the World Health Assembly recognized e-health as the way to achieve cost-effective and secure use of Social Media technologies for health.

The use of Social Media for e-health services can offer important benefits mainly in three different areas:

- Productivity: cost reduction and avoidance, increased productivity, reduced duplication of tests/procedures and impacts on success of reform or change initiative.
- Access: an easier access to health services mainly in remote areas and reduction in wait-times for medical procedures.
- Quality: improving the quality of life; privacy and security incidents and positive audit or operational review observations.

These benefits underline that today the use of the Social Media can be very important to improve the accurate delivery of e-health. Social Media can be used to achieve healthcare goals in four areas: communications, information sharing, clinical outcomes and speed innovation that are very important both for physicians and patients. Physicians have the opportunity of learning, professional development and collaboration among them. Patients can promote wellness and healthy lifestyles, share advice on treatments and provide motivational support among other healthcare consumers. In this context a great importance is having the Cloud computing that allows millions of Internet users that use Social Media platforms to exchange messages, upload and share images, videos and data and collaborate on-line. Cloud computing is a new model of delivering computing resources anytime from anywhere, based on on-demand self-service Internet infrastructure. According to Rosenthal et al. [9], the biomedical informatics community, can take advantage of the new computing paradigm, sharing data and applications. Informatics innovations showed that cloud computing can overcome health data management and analysis issues [10, 11]. It provides three new advantages: massive computing resources available on demand, elimination of an up-front commitment by users and payment for use on a short-term basis as needed [12].

Starting from these considerations, this chapter analyses the e-health services provided by different Social Media (collaborative projects, blogs, content communities, social networking sites, virtual games and virtual social worlds and video-chat) and introduces a Hybrid Cloud E-health Services architecture (HCLES) able to provide open, interoperable, scalable, and extensible services for all the e-health activities. It integrates the potentialities of Skype for a direct communication and synchronous data transmission among people with the cloud perspective of Social Media services.

The chapter is organised as follows. In Sect. 2 an overview of different studies on Social Media in healthcare sector is provided. Section 3 describes the different activities and relationships on Social Media among physicians and patients. In Sect. 4 an analysis of different e-health services provided by different Social Media is carried out. Section 5 introduces the Hybrid Cloud E-health Services architecture (HCLES) on health services. Finally Sect. 6 concludes the chapter.

2 Background

The Web has been transformed from a place with many users and few providers, into a space where everyone can have a voice by tools such as forums and blogs or communities of users that can collaboratively create a body of knowledge about some concrete topics. This evolution produced the development of Web 2.0 or Social Web [13]. The Web 2.0 promises to boost human collaboration capabilities on a worldwide scale, enabling individuals to organise themselves in Social Media in order to share information and services and to collaborate by means of read-write Web and user generated content. As Social Media and in particular Social Networks are becoming popular, many studies are performed to investigate their properties to understand their practices, implications, culture, and meaning. Therefore, many methods are developed to collect [14–15] and to visualise [16–17] network data in order to analyse relationships between people, groups, organisations and other knowledge processing entities on the network. The literature indicates Social Media in healthcare sector as "websites where consumers may be able to find health resources at a number of different levels. Services may range from a basic tier of emotional support and information sharing to questions and answers with physicians to quantified self-tracking to clinical trials access" [18].

This growing phenomenon is called Health 2.0 and is defined as: "the use of social software and its ability to promote collaboration between patients, their caregivers, medical professionals, and other stakeholders in health" [19]. Many studies have been developed in order to understand the reasons that lead people to join Social Media for health purposes.

According to Levy [20] people use Social Media to see what other people say about a medication or treatment, to research other people' health knowledge and experiences, to learn skills or get education that helps to manage a condition and to get emotional support. Similarly, doctors join Social Media to share ideas and debate treatment options, and they can produce a collective decision to improve quality of cares [21].

D'Andrea et al. [22] analysed Italian doctors perceptions about advantages and disadvantage in using Social Media for relationships between patients and physicians, among patients and among physicians. For the first relationship, the main advantage consists in the possibility to share opinions and medical information every time. This sharing process improves the awareness of patient's health conditions enhancing their satisfaction, especially for patients with chronic illnesses and patients living in rural or outlying areas. The main disadvantage underlined in using Social Media was the impossibility of the face to face presence that could sometimes improve the probability of an incorrect diagnosis with the risks of misunderstanding by patients. On considering the relationship among patients, using Social Media usually improves the capacity to provide information and moral support they can give each other, by sharing of diseases and experiences assuring online anonymity. Among risks, patients could increase each other panic toward a particular pathology, stimulating hypochondria, and pass information about the treatments between them, suggesting the administration of dangerous and wrong drugs. With respect to relationship among physicians, the use of Social Media improves their possibilities to participate in continuous medical education along with useful communication and collaboration and knowledge-sharing, accelerating the emergence of trends and new insights on medications, treatments and devices.

Among risks, the privacy issue related to patients data and the danger that multimedia files could be intercepted and used by unauthorized people for non-medical purposes was underlined.

According to a recent survey of 2013 [23] by the Pew Research Center's Internet & American Life Project, 72 % of internet users looked online for health information for minor and serious health problems and general information searches, they are called "online health seekers." Seventy-seven percent of them started to search for health information on search engine such as Google, Bing, or Yahoo, 13 % at a specialized site in health information, like WebMD, 2 % started their research at a more general site like Wikipedia and 1 % at a social network site like Facebook.

According to the research, Internet is used as a diagnostic tool, in fact 59 % of U.S. adults searched online for health information in the past year. The survey found also that people want to have peer-to-peer support, they exchange stories about their own health issues to help each other understand what might lie ahead: 26 % red or watched in the past year someone else's experience about health or medical issues. Sixteen percentage of Internet users went online in the past year to find others who might share the same health concerns. Some studies have shown positive effects of supportive social networks on health and on preoperative pain and anxiety [24]. Kelly Barnes, U.S. Health Industries leader for PwC, said in a statement of 2012 that "the power of Social Media for health organizations is in listening and engaging with consumers on their terms. Social Media has created a new customer service access point where consumers expect an immediate response. Health organizations have an opportunity to use Social Media as a way to better listen, participate in discussions, and engage with consumers in ways that extend their interaction beyond a clinical encounter" (http://searchenginewatch.com/article/2169462/33-of-U.S.-Consumers-Use-Social-Media-for-Health-Care-Info-Survey).

A research of EPG Health Media [25] found that UK healthcare professionals use Social Media for healthcare purposes more than the other countries. France and Germany, instead have a biggest percentage of "never" in engaging in Social Media respect to the other countries. Figure 1 shows, in detail, how many time healthcare professionals engage in health-related discussions via online Social Media in big five European markets.

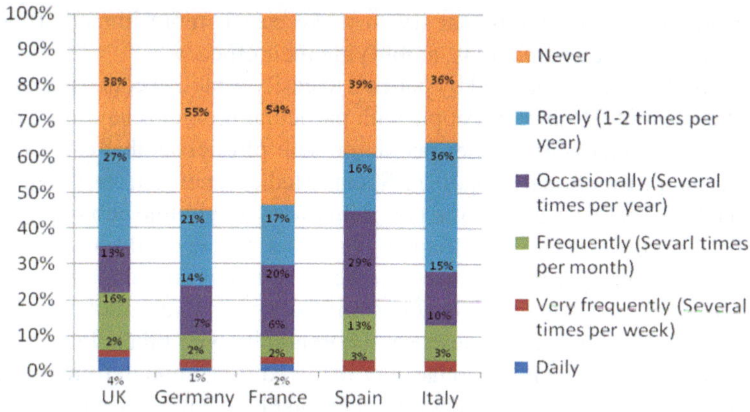

Fig. 1 Use of Social Media by physicians

3 Social Media: Interactive Group Activities Between Patients and Physicians

Social Media in e-health sector are socially helpful because they allow people to easily establish contacts among them and to have a sounder mutual knowledge.

In Social Media people can be organised according to three main kinds of groups:

- Physicians who actively seek each other and share authentic information and knowledge;
- Patients that share common health problems (e.g. cancer, infertility, neurodegenerative diseases, etc.) giving and receiving emotional and psychological support;
- Physicians and patients that share opinions and medical information.

In this perspective, the widespread diffusion of Social Media allows physicians to participate in continuous medical education at a time and location convenient for them, along with:

- Useful communication and collaboration
- Knowledge-sharing

Physicians aggregate observations from their daily practice and challenge or collaborate each others' opinions using a virtual space, accelerating the emergence of trends and new insights on medications, treatments and devices [26]. As consequence, it emerges the increasing number of skills, competencies and "knowledge profiles" of each physician involved in the Social Media. In this perspective, Social Media amplify openness, interoperability, scalability, and extensibility of a traditional community. At the same time, this does not mean that in Social Media face-to-face interactions are substituted by electronic communications. On the contrary, face-to-face interactions among physicians are enabled and supported by electronic communications that try to make easier, more immediate and less expensive the knowledge relations, amplifying their efficacy.

Social Media can also be used by patients to easily establish contacts among each other to have a mutual support. People with particular problems have the possibility to share their problems with other people or with physicians that can facilitate to overcome them. Emotional support in Social Media benefits from the absence of traditional barriers to access and the possibility to assure online anonymity that can be helpful for those who have stigmatising or embarrassing conditions. These forms of emotional support help to reduce members' isolation and empathy and, at the same time, they contribute in enhancing self-esteem and sustaining hope. People join Social Media to experience a sense of community with others like themselves. To obtain this benefit, members must stay involved long enough to feel a sense of connection with other members. They are more likely to have this experience if they begin exchanging messages with other members.

Social Media are rapidly changing also the physician–patient relationship by allowing them to share opinions and medical information every time and everywhere. Social Media offer to physicians an opportunity to improve the awareness of patient's health conditions and enhance their satisfaction. They also give the opportunity to increase the involvement of patients in their treatments and they improve access to health care information and communication possibilities between patients and physicians. Through the use of Social Media physicians have the possibility to communicate with patients continuously allowing a more accurate data collection [27]. Social Media can be used by doctors to send instant messages to patients reminding them when they need to take their medication, reducing certain administrative costs associated with hospitalisation that are a consequence of incorrect patient's behaviors in following the prescribed medication at the correct time. There are many potential benefits for patients and physicians who communicate in Social Media for instance patients may feel more comfortable in addressing sensitive, complex or personal issues; moreover Social Media can solve problems related to large distances or patients' disability.

However Social Media does not have the capability to reproduce the traditional relationship because of the impossibility of the physical presence. In fact medical practice includes complex processes as diagnosis, treatment, prognosis and these processes require the presence of the patient for several activities. Social Media can modify and integrate the traditional physician–patient relationship but, at the moment, cannot replace this relationship. In order to incorporate virtual consultations into routine medical practice it is necessary to proceed on the basis of secure evidence. It is important to understand the following aspects related to the communication between physicians and patients: (1) how the communication by Social Media can be integrated with other modes to communicate; (2) which are the patient's and physician's preferences in using Social Media; (3) how to identify people that most likely can benefit from virtual communication.

4 Social Media e-Health Services

There are various types of Social Media used by patients and physicians. Kaplan and Haenlein [28] define Social Media as "groups of Internet-based applications that build on the ideological and technological foundations of Web 2.0 and that allow the creation and exchange of user-generated content". In particular the authors provide six different classes of Social Media: collaborative projects, blogs, content communities, social networking sites, virtual games and virtual social worlds. To this classification we added also the video-chat as they are increasing their presence in e-health mainly for the on-line therapy. In Table 1 the different e-health services provided by the different Social Media are summarized.

On considering *Blogs*, these are diaries of text or pictures in chronological order (such as Blog spot) where contents are more static and includes every kind of multimedia items (such as MySpace) or every other thematic social contents (list of Web links, list of news, musical tastes, and so on). While initially blogs were personal and published under pseudonyms, today their development led to professional and expert blogs and they are become an accepted method of journalism. Blogs can be written by formal organizations like businesses, non-profits or even governments, or by individuals. A blog may be shared by two or more people with a common interest or can be an individual's effort on thoughts, experiences and advices. A health blog is a blog that focuses on medical, preventive, experiential, political or other health-related topic. Blogs are used by the medical community, in particular by trainees or practicing physicians that write on clinical practice and suggestions on health [29]. This facilitates the access of patients to expert opinions, but at the same time this can cause some threats for the traditional doctor–patient relationship and for ethical issues such as breaches of patient confidentiality [30]. There are blogs both for patients and physicians. A typical example of blog for patients, is (http://patientadvocare.blogspot.it/) in which patients can build a community, helping someone else, teaching a lesson, bearing witness about patient safety, medical malpractice and preventing future errors. While examples of blogs for physicians, are (http://drmichellecleere.com/blog/) where medical professionals have the possibility to share opinions and (http://doc2doc.bmj.com/) where they can give medical support to patients.

Collaborative projects are the joint development of contents to be published on a website. The greatest example is Wiki, a web-based application that allows users to add content and also to edit content supporting collaborative writing, opening discussions, interaction and web-authoring [31]. At the same time it also allows to delete the content of a new page or another already written by others, to restore an old version, to avoid risks of damages by hackers or non-collaborative members. Thanks to wiki everybody can write everything about a topic and can read everything, or improve what is already written. Wiki is the most important example of collaborative online community, and applied to a healthcare site gives to every patient the chance to share their own experience and to collaborate with other members, activeness and loyalty to the site is guaranteed. An example of

Table 1 Social Media classification

Social media	Usage	E-health services	Example
Blog	Patients	Receive and give emotional and psychological support	http://patientadvocare.blogspot.it/
	Physicians	Give medical support to patients	http://drmichellecleere.com/blog/
		Share opinions with others physicians	http://doc2doc.bmj.com/
Collaborative projects	Patients	Access to medically-related information	http://www.wikihealth.com/Main_Page
	Physicians	Provide medically-related information	
Content communities	Patients	Share videos on health	http://www.youtube.com/education?category=University/Medicine
	Physicians		
Social networking sites	Patients	Share experiences	www.patientlikeme.com
	Physicians	Offer advice about clinical questions to patients	www.medgle.com
		Share knowledge with others physicians	www.doctors.net.uk
Virtual gaming worlds	Patients	Improve healthcare behavior	Heartlands
	Physicians	Transfer skill from virtual to real settings	
Virtual social worlds	Patients	Receive and give emotional and psychological support	http://slurl.com/secondlife/SupportforHealing/111/166/53
	Physicians	Give healthcare education to patients and others physicians	http://slurl.com/secondlife/Eduisland%20II/203/21/22
Video-chats	Patients	Receive on-line therapy and counseling	www.skype.it
	Physicians	Give health consultation to patients and others physicians	

a collaborative and updated online health and wellness community is the website WikiHealth http://www.wikihealth.com/Main_Page. Its goal is to offer the most comprehensive, current and insightful information to help anyone and everyone achieves optimal health. People knowledge can help others by sharing what they know. Another example is http://en.wikipedia.org/wiki/Portal:Medicine, used both by patients to access to medically-related information and physicians to provide medically-related information. A feature of a wiki is to make changes with minimal effort by any individual, it can cause some concerns about inaccurate information and, for this reason some wikis (with medical information) restrict content changes to registered users. Furthermore bad information is quickly identified and corrected.

Content communities are online databases of multimedia content where users can search contents for information about particular topics. The main objective of this Social Media is to provide content sharing among users. An increasing amount of content is disseminated on social video platforms, such as YouTube. More than one billion unique users visit YouTube each month and approximately 100 million users take social action on YouTube (such as, shares, likes, and comments) every week (http://www.youtube.com/t/press_statistics). YouTube is gaining popularity in different sectors; one of this is represented by the e-health. Different studies have been carried out to analyse health video on YouTube in particular on anorexia, rheumatoid arthritis, prostate cancer, cardial infarction, H1N1 virus, papillomavirus vaccine, immunization [32–38]. Results of these studies shown that among American and European physicians and patients YouTube is used "not just as a video repository, but also as a social network where users interact to build trust with comments and favorites" [39]. According to Chou et al. [40], patients mainly publish videos about their diseases whereas physicians collaborate with video sharing to increase the quality of health knowledge.

A *social networking* site is an online service, which focuses on facilitating the building of social relations among people who share interests, backgrounds, activities, or real-life connection. Social Networks have the great potential to serve ubiquitous information and communication needs. This applies especially to the healthcare sector where the need for information, communication and services is deeply felt. Social networks in e-health sectors allow patients to share health experiences and receive advices about clinical questions and physicians to share knowledge with others physicians. A typical example of Social Network community of patients is Patientlikeme (www.patientlikeme.com). On PatientsLikeMe's network, people connect with others who have the same disease or condition and track and share their own experiences. In the process, they generate data about the real-world nature of disease that help researchers, pharmaceutical companies, regulators, providers and nonprofits to develop more effective products, services and care. On considering physicians, an example of Social Network community used by them is Doctors.net.uk. This represents one of the largest and most active networks of medical professionals in the UK, it is available to UK-registered doctors in primary and secondary care, as source of medical education, research and communication. It offers a professional, secure e-mail facility, clinical and non-clinical data, the latest medical information and free accredited education allowing doctors to maintain Continuing Professional Development (CPD). While an example of Social network site used by physicians to offer advice about clinical questions to patients is MedHelp (http://www.medhelp.org/). Through MedHelp's patients, doctors, experts, researchers, hospitals, and non-profit organizations, work together to find cures together. Every day, members come to MedHelp to receive the support they need from other patients like them, to research information on drugs and health topics, and to share their knowledge with others.

A *virtual game* is a three dimensional environment where users can interact using a personal avatar. The high popularity of this Social Media represents a focal topic in different research areas. There are a wider range of online games: casual games, serious games and advergames. Casual games are purely for entertainment while serious games are designed for improving specific aspects of learning. Finally advergames use the technique of persuasion to promote a brand, product or a political candidate. On considering the e-health sector, the most used games to educate patients on health topics are the serious games. According to Roubidoux et al. [41] "since 2002 many serious games in the e-health sector have been developed, dealing with a wide variety of aspects of surgeon training, radiology operation, Car-diopulmonary Resuscitation (CPR) and patient care, among others". Serious games have had a big impact on the health sector because they help patients to improve their healthcare behavior and physicians to transfer skills from virtual to real settings. An example of a serious game in e-health sector is Heartlands. This virtual game allows to maps physical movement by using a GPS-equipped smartphone attached to a heart rate monitor. The game allows patients to achieve a personal best, sport activity with a personal gaming challenge and researchers to understand the little explored area of discontinuous exercise.

Virtual social worlds allow people to live a "virtual life" in parallel to their real life. The most famous example is Second Life, an online virtual world where users can interact with each other through avatars. In recent years, professionals and universities are using it as an educational platform to improve training in many different areas including the healthcare sector. An example of its use in healthcare education is at Imperial College London. They have built a virtual hospital where students can see patients, order X-rays, consult with colleagues and make diagnoses 24 h a day [42]. Another example is from the University of North Carolina's (UNC) School of Pharmacy. Students can play the pharmacist and take a quiz inside of the virtual environment to become comfortable with the places that they will be going to in real life. It is possible also to organise medical conferences that in real world are very expansive. Physicians can also give some consultations to patients, for example the "Counseling Center", provides counseling and consulting on health issues. Patients can also learn on Second Life, one example is the "Nutrition Game" where patients can learn more about nutrients and the negative effect of fast food on health. Second Life also provide patients with opportunities regarding support centers for people dealing with health problems. Examples are given by the American Cancer Society, Red Cross and Alcoholics Anonymous and the Heron Sanctuary, a support community that helps patients to find answers.

Video-chats are social media used for real time audio/video interactions between users at disparate locations [43]. They are typically conducted via computer, smart-phone device or tablet and may involve or one-to-one or one-to-many interactions. During a video/audio conferences, participants can see whatever is on the presenter's screen, and simultaneously share applications and discuss matters of common topic. The combination of video data with voice interaction makes possible to (1) collaboratively perform activities in different parts of the world; (2) provide mentoring or homework help. One of the most used video-chat that is having a particular relevance in the e-health sector is Skype. Skype is becoming increasingly popular mainly for online therapy and counseling. Skype creates an interactive environment that stimulates the collaboration and interaction between patients and physicians, and among physicians. Although it is unlikely to completely replace traditional face to face counseling many are the benefits of the Skype on-line therapy. First of all, Skype allows patients a greater sense of privacy while still offers a face-to-face interaction if desired. Moreover it allows physicians to reduce costs and to provide services to patients who could not normally receive for their physical impairment or their isolation.

The use of Social Media, jointly with on-line services both by patients and physicians in the emerging perspective of cloud computing provides a more flexible and scalable approach. In this perspective we propose an Hybrid Cloud E-health Services (HCLES) architecture, integrating the potentialities of Skype for a direct communication between people with the cloud perspective of Social Media services. In the following section this architecture is described in detail.

5 An Hybrid Cloud E-Health Services Architecture (HCLES)

Starting from the analysis of the different uses of Social Media for e-health and considering the need to provide open, interoperable, scalable, and extensible services for all the e-health activities, a Hybrid Cloud E-health Services architecture (HCLES) is proposed. The architecture integrates the potentialities of Skype for a direct communication between patients and physicians with the cloud perspective of Social Media services (as shown in Fig. 2).

The proposed architecture, with its hybrid solution, proposes to use Skype for services implying synchronous data transmission (for example tele-consulting), while the components of the cloud platform provide synchronous communication activities among the different actors of Social Media.

The Skype user directory is entirely decentralised and distributed among the nodes in the network. This implements scalability without a complex infrastructure. Each physician connects himself/herself with the others accessing, annotating or adding comments at the patient data using multimodal input/output channels. The telemedicine centre of Shanghai Hospital is the centre handling the second largest number of tele-consultations in the entire network. In Ping [44], the author examined the medical records from telemedicine cases dealt with this telemedicine centre. Authors underline that three aspects of the management of the medical records, with a particular attention on tele-consulting, are particularly important: multimedia collection, standardization of patient/record identification and classification, and information management. Multimedia collection is particularly important for Patient monitoring and Healthcare-learning too. In fact, during tele-consultation as well as Healthcare-learning activities the need to share multimedia data, and

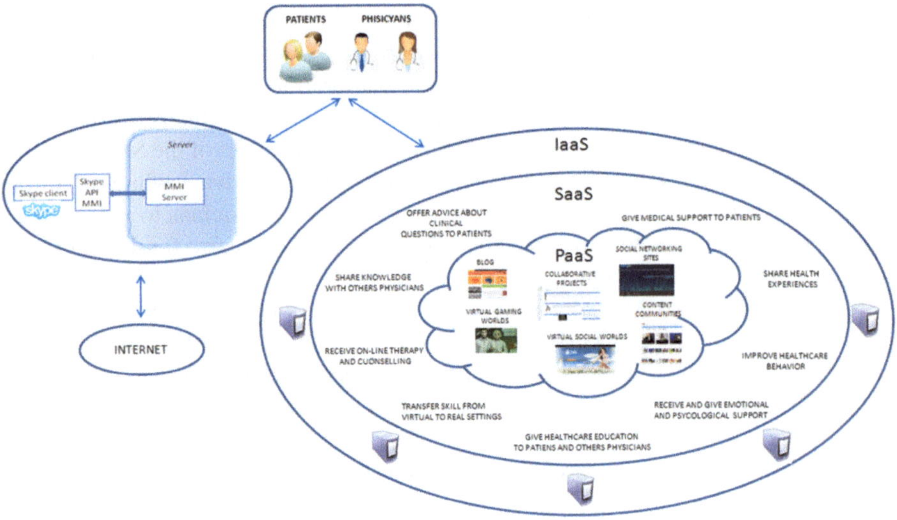

Fig. 2 The Hybrid Cloud E-health Services architecture (HCLES)

facilities that can emerge by using multimodal interaction channels suggested us to define an architecture supporting multimodal interaction in social networking. This is the situation when general physicians and specialists organize themselves to exchange professional information and opinions, improving their personal and collective knowledge and sharing wide clinical occurrences. The Skype platform and peer-to-peer communication tools allow a wide sharing of data and knowledge, implementing services for tele-consulting, and an interactive and continuous learning of participants. Privacy data need to be managed. The registered physicians have privileges to access and interact with clinical data such as (RX, RMN, etc.).

On considering the cloud platform, this consists of:

- Infrastructure as a Service (IaaS): that involves hardware (servers, storage, network technologies,). It provides a Web interface that allows patients and physicians accessing virtual machines.
- Software as a Service (SaaS): includes business applications hosted and delivered as a service via the Web that do not require installation of additional computer programs. On this component there are all the e-health services offered by Social Media to patients and physicians.
- Platform as a Service (PaaS): offers an integrated set of Social Media that provides an online environment for a quick communication and collaboration between patients and physicians.

Integrating these two approaches (Skype and cloud platform) correspond to the need of supporting existing communities with services by Skype, facilitating their connection, and creating communities of interests of patients, physicians or hybrid communities that, provide/receive emotional and psychological support, medical support, medical information, health care education, and tele-consulting.

With respect to security and privacy issues these involve information about people through spatial and temporal extension of data collection activities. Many studies in the literature are predominately focused on using different forms of access control mechanisms for prevention and protection. Juels [45] provided a technical solution on the problems of privacy and security through the cryptographic mechanisms and symmetric-key tags approaches.

According to these aims the platform will collect sensitive personal information about the users by providing the following technical solutions to increase their privacy and security:

- Authentication mechanisms: to verify the users which data is coming from. In order to provide the authentication mechanisms different authentication algorithms such as digital signatures, passwords and challenge response authentication protocols, will be used.
- Encryption: to guarantee the security of the data and to prevent skimming and eavesdropping. In order to provide encryption in software, symmetric and asymmetric key algorithms provided in [46] will be used.

6 Conclusion

The chapter described e-health services provided by different Social Media such as (collaborative projects, blogs, content communities, social networking sites, virtual games and virtual social worlds and video-chat) and introduced a Hybrid Cloud E-health Services architecture (HCLES) able to provide open, interoperable, scalable, and extensible services for all the e-health activities.

The proposed architecture integrated the use of Skype for services implying synchronous data transmission (for example tele-consulting) and the cloud platform that provides synchronous communication activities among the different actors of Social Media.

The Skype platform allows a wide sharing of data and knowledge, implementing services for tele-consulting, and an interactive and continuous learning of participants. While the cloud architecture consists of: Infrastructure as a Service (IaaS) that provides a Web interface that allows patients and physicians accessing virtual machines. Software as a Service (SaaS): that includes all the e-health services offered by Social Media to patients and physicians and Platform as a Service (PaaS): offers an integrated set of Social Media that provides an online environment for quick communication and collaboration between patients and physicians.

Future works will involve the validation of the architecture both in terms of interaction and clinical services. Moreover following the next technological evolution it could be possible to integrate also synchronous communication in the cloud architecture.

Glossary

Blog Is an Internet discussion site that consist of discussion posts.

Cloud-Computing Is a term used to describe different types of computing concepts that involve various computers connected through Internet.

E-health Indicates healthcare practice supported by electronic communication and processes.

Healthcare Is the prevention and treatment of illness, disease, injury, and other mental and physical impairments in human beings.

Social Media Refers to the means of interactions in which people share, create and exchange information in virtual communities.

Social Networking Is group of people that discuss and share ideas and/or specific interests on the Internet.

References

1. Vuori M (2012) Exploring uses of social media in a global corporation. J Syst and Inform Technol 14(2)
2. Moran M, Jeff S, Hester T-K (2011) Teaching, Learning, and Sharing: How Today's Higher Education Faculty Use Social Media. Babson Survey Research Group
3. Ferri F, Grifoni P, Guzzo T (2012) New forms of social and professional digital relationships: the case of Facebook. Social Network Analysis and Mining Journal. vol. 2. Springer, pp 121–137
4. Wang Y, Yu Q, Fesenmaier DR (2008) Defining the Virtual Tourist Community: Implications for Tourism Marketing. Tourism Manage 23(4):407–172001
5. Eysenback G (2001) What is e-health? J Med Internet Res 3(2):e20
6. Marconi J (2002) E-health: navigating the internet for health information healthcare. Advocacy white paper. Healthcare Inf Manage Syst Soc
7. Wyatt JC, Liu JLY (2002) Basic concepts in medical informatics. J Epidemiol Community Health 56:808–812
8. Akeh L, Morfaw Z (2007) E-health Africa: overcoming the barriers to its implementation. A Case Study of Sub Sahara Africa. Second Annual ICT4D Postgraduate Symposium (IX): e-Health and eGovernment
9. Rosenthal A, Mork P, Li MH, Stanford J, Koester D, Reynolds P (2010) Cloud computing: a new business paradigm for biomedical information sharing. J Biomed Inform 43:342–353
10. Dudley JT, Butte AJ (2010) In silico research in the era of cloud computing. Nat Biotechnol 28(11):1181–1185
11. Botts N, Thoms B, Noamani A, Horan TA (2010) Cloud computing architectures for the underserved: public health cyber infrastructures through a network of health ATMs. In: Proceedings of the 43rd Hawaii international conference on system sciences, pp 1–10
12. Armbrust M, Fox A, Griffith R, Joseph AD, Katz R, Konwinski A, Lee G, Patterson D, Rabkin A, Stoica I, Zaharia M (2010) A view of cloud computing. Commun ACM 53(4):50–58
13. Reilly T (2005) What is web 2.0. Design patterns and business models for the next generation of software. Available at http://www.oreillynet.com/pub/a/oreilly/tim/news/2005/09/30/what-is-web-20.html
14. D'Andrea A, Ferri F, Grifoni P (2009) An overview of methods for virtual social network analysis. In: Abraham, Ajith et al. Computational Social Network Analysis: Trends, Tools and Research Advances. Springer, http://books.google.com/books?id=-S1KiURSfRAC&pg=PA8

15. Hanneman RA, Riddle M (2011) Concepts and measures for basic network analysis. The sage handbook of social network analysis. Sage. pp 346–347. http://books.google.com/books?id=2chSmLzClXgC&pg=PA346
16. Heer J, Boyd D (2005) Vizster: visualizing online social networks. IEEE symposium on information visualization (Info Vis 2005), Minneapolis, MN, USA, 23–25 October
17. Caschera MC, Ferri F, Grifoni P, Guzzo T (2009) Multidimensional visualization system for travel social networks. 6th International conference on information technology: new generations ITNG 2009, IEEE computer society, Las Vegas, April 27–29, pp 1511–1516
18. Swan M (2009) Emerging patient-driven health care models: an examination of health social networks, consumer personalized medicine and quantified self-tracking. Int J Environ Res Public Health 6:492–525, Available at: http://www.mdpi.com/1660-4601/6/2/492/
19. Sarasohn-Kahn J (2008) THINK-Health. The wisdom of patients: health care meets online social media. California HealthCare Foundation
20. Levy M (2007) Online health: assessing the risk and opportunity of social and one-to one media. Jupiter Research
21. Surowiecki J (2005) The wisdom of crowds. Anchor Books, New York
22. D'Andrea A, Ferri F, Grifoni P, Guzzo T (2010) Multimodal social networking for healthcare professionals. In: Proceeding of the fourth international workshop on management and interaction with multimodal information content (MIMIC'10), 30 August–3 September 2010, Bilbao. IEEE Computer Society Publishing
23. Pew Research Center's Internet & American Life Project, Health Online (2013) California Healthcare Foundation (January 15, 2013)
24. Mitchinson AR, Kim HM, Geisser M, Rosemberg JM, Hinshaw DB (2008) Social connectedness and patient recovery after major operations. J Am College Surgeons 206(2):292–300
25. EPG Health Media, Social Media and Healthcare: How do Healthcare Professionals (2010) Patients/Consumers and Pharmaceutical Companies use social media in relation to health? Market research
26. Ebner W, Leimeister JM, Krcmar H (2004) Trust in virtual healthcare communities: design and implementation of trust-enabling functionalities. In: Hawaii international conference on system sciences (HICSS), Big Island
27. Leimeister JM, Krcmar H (2006) Designing and implementing virtual patient support communities: A german case study. In: Murero M, & Rice RE, (eds) The internet and health care: Theory, research and practice. Verlag: Lawrence Erlbaum Associates; Erscheinungsort. Mahwah: Erscheinungsjahr
28. Kaplan AM, Haenlein M (2010) Users of the world, unite! The challenges and opportunities of social media. Bus Horizons 53(1):59–68
29. Emory University. Student Life Blogs: The Second Opinion (2011). Available from: http://www.med.emory.edu/blog/. Accessed 9 Sept 2011
30. Dainton C (2009) Physician-writers in the age of blogging. CMAJ 181(5):348
31. Desilets A, Paquet S, Vinson N (2005) Are wikis usable?. In: WikiSym conference, San Diego, October 16–18
32. Syed-Abdul S, Fernandez-Luque L, Wen-Shan J, Yu-Chuan L, Crain S, Min-Huei H, Yao-Chin W, Khandregzen D, Chuluunbaatar E, Anh Nguyen P, Der-Ming L (2012) Tweetations for "Misleading Health-Related Information Promoted Through Video-Based Social Media: Anorexia on YouTube". J Med Internet Res 15(2):e30
33. Singh AG, Singh S, Singh PP (2012) YouTube for information on rheumatoid arthritis—a wakeup call? J Rheumatol 39(5):899–903
34. Steinberg PL, Wason S, Stern JM, Deters L, Kowal B, Seigne J (2010) YouTube as source of prostate cancer information. Urology 75(3):619–622
35. Pant S, Deshmukh A, Murugiah K, Kumar G, Sachdeva R, Mehta JL (2012) Assessing the credibility of the "YouTube approach" to health information on acute myocardial infarction. Clin Cardiol 35(5):281–285
36. Pandey A, Patni N, Singh M, Sood A, Singh G (2010) YouTube as a source of information on the H1N1 influenza pandemic. Am J Prev Med 38(3):e1–e3

37. Ache KA, Wallace LS (2008) Human papillomavirus vaccination coverage on YouTube. Am J Prev Med 35(4):389–392
38. Keelan J, Pavri-Garcia V, Tomlinson G, Wilson K (2007) YouTube as a source of information on immunization: a content analysis. JAMA 298(21):2482–2484
39. Fernandez-Luque L, Karlsen R, Melton GB (2012) HealthTrust: a social network approach for retrieving online health videos. J Med Internet Res 14(1):e22
40. Chou W, Hunt Y, Folkers A, Augustson E (2011) Cancer survivorship in the age of YouTube and social media: a narrative analysis. J Med Internet Res 13(1):e7
41. Roubidoux MA, Chapman CM, Piontek ME (2009) Development and evaluation of an interactive web-based breast imaging game for medical students. Acad Radiol 9:1169–1178
42. Bradley J (2009) Can second life help doctors to treat patients? In: CNN. Retrieved from http://www.cnn.com/2009/TECH/03/30/doctors.second.life/
43. Lensegrav P, Pearce K (2002) The responsiveness of elementary students to the use of video conferencing. Retrieved from http://www.bhsu.edu/education/edfaculty/lq>earce/Responsiveness%20ofl'ib20Elementarv%20Students%20to%20Video%20Conferencing.htm
44. Ping L (2003) The quality of medical records in teleconsultation. J Telemed Telecare 9:35–41
45. Juels A (2006) RFID security and privacy: a research survey. IEEE J Selected Areas Commun 24(2):381–394
46. Meingast M, Roosta T, Sastry S (2006) Security and privacy issues with health care information technology. In: Conference proceedings of the IEEE engineering in medicine and biology society, vol 1, pp 5453–5458

Index

Ş. Gündüz-Öğüdücü and A.Ş. Etaner-Uyar (eds.), *Social Networks: Analysis and Case Studies*, Lecture Notes in Social Networks, DOI 10.1007/978-3-7091-1797-2,